Prostitutes, Margarine and Handguns

ALBERT D. MCCALLUM

Pragmatic Publications
Springport, Michigan

Pragmatic Publications
18440 29½ Mile Road
Springport, Michigan 49284

Library of Congress Catalogue Card Number: 96-92098

ISBN: 0-9651659-0-6

First printing May 1996

Printed in the U.S.A.

I dedicate this book to my mother, Evelyn McCallum,
and the memory of my father, Homer McCallum,
who provided an uncommon environment
for my formative years.
Without this perspective I might never have seen
the world of which I write.

ACKNOWLEDGEMENTS

Every book is the work of a lifetime. Every person who touches the mind of the author in some way influences his work. Relatives, teachers, friends, coworkers and others who crossed my path, many whom I could name and many whose names I don't recall or never knew, have in some way contributed to the expressions transcribed on the pages which follow. The intangible contributions are far too many to list.

Some have made recent and tangible contributions. Bob Doner, publisher and editor of *The Springport Signal* and Tom Nolte, publisher and editor of the *Times of Grass Lake/Waterloo* were the first to publish my column. Without these opportunities to present my work to the public and to hone my writing skills, it is most unlikely that this book would now be in print.

My wife, Arlona, proof read the manuscript several times as well as listening to my thoughts, ramblings and observations before they reached the printed page.

Feed back, both written and oral, from the readers of *Thoughts, Ramblings and Observations* contributed immensely to my decision to go ahead with a book.

I wish to express my appreciation to all of those who have contributed to the forming and recording of the reflections I express on the following pages.

Albert D. McCallum

CONTENTS

Contents

Contents

INTRODUCTION

We are constantly reminded of the complexity of our world. Certainly the technology surrounding us is complicated. We must not forget that the main ingredient in any civilization is people. I would not suggest that people are not complex. Complex as we may be, might I suggest that we are no more complex than our ancestors in pioneer days.

The crucial problems we face as individuals, and as a nation, involve people, not machines. If our ancestors could maintain stable relationships with each other a hundred years ago, Why can't we today?

Might it be that we are so enamored with our technical gadgetry that we overlook the little human factors that glue a civilization together – and provide the explosives that can blow it apart? In the following pages we will explore some of the myths, illusions and delusions that obscure the human factors which hold the keys to our future. Occasional we will journey to the fountain of frivolity to quench any thirst for the lighter side.

Might we discover that, like with a thousand-piece puzzle, we have many small problems, not a few big ones. That would be good news. Might we, the people, be able to solve little problems instead of waiting for the government bus to come and take us all to utopia? The government bus can be slow in coming. What should we do when it's headed in the wrong direction? Must we pay whatever fare the driver demands?

Albert D. McCallum

OUR NATIONAL SYMBOL

The eagle did not run unopposed for selection as our national symbol. Ben Franklin nominated the turkey. Don't be too hasty to conclude that old Ben's best years must have been behind him. The turkey of Ben's day wasn't quite the bird we put on the Thanksgiving table. Those clumsy birds that can't even breed by themselves were unknown in the eighteenth century.

The bird Ben had in mind was more akin to the wild turkeys now roaming much of my home state of Michigan and many other states. If you have seen these wily birds in their habitat, you know they aren't . . . they just aren't . . . I guess what I'm trying to say is, it's an injustice to call such noble birds turkeys.

Others have questioned the wisdom of making the eagle our national symbol. The soaring eagle is a majestic sight. The bird is strong. It's powerful. It's quick. The eagle is also a vicious predator. It is known to eat carrion (dead stuff). Of course, vicious predator is not a characteristic unknown in the human species. Humans also commonly eat dead stuff. The humans who eat live stuff are the ones who seem strange. Most of us are too young to have witnessed that passing fad of swallowing live gold fish. Remember Ozzy Osbourne and the chicken heads?

> **It would be politically incorrect to discriminate against the ant because of the form in which it was created. The ant has as much right to be our national bird as you do.**

I'm not concerned about the merit, or the lack of merit, of turkeys and eagles. I do, however, propose to change our national bird. What is my nomination? The ant. Someone is bound to point out, the ant isn't even a bird. It isn't the ant's fault it's not a bird. It would be politically incorrect to discriminate against the ant because of the form in which it was created. The ant has as much right to be the national bird as you do.

Why did I choose the ant? I bet you never guessed I was going to get to that. Ants are industrious. They work together. They get the job done. Not only that. They tackle what would seem an impossible job if they stopped to worry about it. Instead of fretting about the difficulties, the ant grabs a grain of sand and is on its way. The ant doesn't stop until the job is finished. Of course, the ant's job is never finished.

Our nation needs more ants. The tasks before us are enormous. No eagle is going to swoop down out of the sky and carry away our problems. Even the great eagles that nest on the banks of the Potomac River don't have that kind of power. Those Potomac eagles claim they have all sorts of powers. You may have noticed, their performances don't live up to their claims. We need millions of ants, each carrying a grain of sand. It is the only way we will ever do the job.

Why do we need the ant as our symbol? When all we can do is drag a grain of sand, it's easy to get discouraged. It's easy to give up. If everybody drops that grain of sand there is no hope for a better future. In fact, there will be no hope that the future won't grow even darker than the present.

Just like with the ant, our jobs will never be done. This is not reason for despair. Mowing the lawn, washing dishes, shoveling snow, cleaning the house, and many other tasks are never done. What would happen if we, or the ant, shunned all the tasks that can't be finished?

From the edge of despair we need to be able to look up and see the symbol of the ant. It will serve as a reminder of the importance of moving that grain of sand. It will remind us that there is no other way. The ant can inspire us all to do our part to build a better America.

The Problem With Pork

If you believe the newscasters and pundits, pork is our most troublesome meat. Not all pork gives us indigestion. The kind kept in barrels causes the problem.

In a strange village on the banks of the Potomac River there resides a tribe of transient, wanabee farmers. Their agricultural abilities are questionable at best. Usually about the only thing they can raise is a disturbance. Give credit where credit is due. They are darn good at raising disturbances.

The truth is, pork is a magical meat. It only looks and smells like pork when it's on someone else's table. When on the voters' own tables it looks like bread and butter.

Semiannually the tribe performs the most sacred ritual passed down by their forbearers. Properly preformed the ritual wards off evil voters, thus ensuring a long and prosperous career. This ritual is, of course, the filling of the pork barrels. The warrior who fills the most and largest barrels is the envy of the entire tribe.

Some less able members of the tribe struggle to fill even a small keg with that magical meat. These members often become rogues. They live at the edge of the village, sniping at their more successful brethren as they load their barrels to send home as peace offerings to their constituents. Voters often mistake this sniping as being motivated by a dislike for pork. In reality it's often nothing more than a king size case of pork envy. It is the new American way. The "have nots" attack the "haves."

How did this quaint ritual begin? Long, long ago voters across the land selected one of their own to venture to the banks of the Potomac to form the tribe now known as Congress. The chosen ones found life could be good in the new village. They wanted to stay. There was a problem. Every two or six years the tribesmen had to go home and plead with their fellow citizens for renewal of tribal membership.

Word of the good life on the Potomac reached home. Upstarts attempted to divest the tribesmen of their birthrights. Sometimes the people back home just didn't seem to care much about renewing the memberships in the Potomac Tribe. Something had to be done to change that. The voters must care.

The more ingenuous members of the tribe started shipping barrels of pork home to their constituents. When the time inevitably came for membership renewal, the savvy tribesmen pointed out that only renewal of their memberships could assure the uninterrupted flow of pork. These tribesmen were not above suggesting that a little gratitude for pork past would be in order. The pitch worked. Memberships were renewed.

Don't blame congressmen for pork barrel politics. They are only giving in to that most basic of human instincts. They do that which is in their own best interest. How many voters would be above filling a few pork barrels if their jobs depended on it?

If it ain't broke, don't fix it. Just perfect it. Over the years the tribesmen fine tuned the delivery of pork to the constituents back home. Pork barrel politics grew to an art form. Constituents sometimes saw juicy roasts and hams delivered to neighboring districts, while their own shipments contained only rinds and shanks. This could lead to the revocation of a tribesman's membership. The example was not lost on the survivors.

Pork turned out to be the most fiendishly addictive substance ever discovered. The survival of the tribesmen depended on their ability to satisfy the appetites of their constituents. This gave rise to an ever expanding flow of pork to the addicts back home.

Some of those people back home complained about the cost of pork. For some strange reason the complaints invariably involved the cost of pork being sent somewhere else. Few, if any, complained about their own pork, or its cost. It was just compensation for all that pork being shipped elsewhere.

The truth is, pork is a magical meat. It only looks and smells like pork when it's on someone else's table. When on the voters' own tables it looks like bread and butter. Constituents find no hypocrisy in demanding their own bread and butter while decrying the wasteful shipment of pork to other communities.

For so long as pork retains that magic, the pork barrels will roll. Until the recipients of the pork recognize it as pork, the demand for pork will continue. For so long as there is a demand, there will be an ample supply. Until voters recognize their own pork and fire the congressmen that ship it to them, pork will prosper. We can fire only the congressman that ships us our pork. We can only complain about the rest. Congressmen could not care less what other voters think of them, as long as their own constituents are content.

Don't blame congressmen for pork barrel politics. They are only giving in to that most basic of human instincts. They do that which is in their own best interest. How many voters would be above filling a few pork barrels if their jobs depended on it? For so long as it serves a congressman's interest to ship barrels of pork, the shipments will continue. The voters best cast the hams out of their own eyes before attempting to cast the chop out of the eye of a congressman.

What Was It?

Something on television the other night surprised me. Just when I thought there was no possible way left for television to ever again surprise me. Had I fallen into a time machine? The show was in color. It couldn't have been a rerun from the fifties. There it was, big as life. Perhaps I should say small as life. What was it? It looked like an Isetta.

Unless you were born before the fifties – unless you have a fantastic memory for the insignificant – you are probably saying, What is an Isetta? I may have misspelled Isetta. It isn't a word that can be found in just any dictionary. I think Isetta rhymes with poinsettia, depending, of course, on how you pronounce poinsettia.

I first laid eyes on an Isetta in the spring of 1957 while visiting the campus of Michigan State University. An Isetta came rolling down the street. It was on the street. It had people in it. It was moving. These are all characteristics of an automobile. The resemblance ended there.

That third wheel was important. Without it the Isetta would have been a motorcycle with a cab. Taking away that third wheel would take away the Isetta's carhood.

Some cars have four wheels and four doors. The Isetta also had four – if you added up the wheels and door. That third wheel was important. Without it the Isetta would have been a motorcycle with a cab. Taking away that third wheel would take away the Isetta's carhood. Actually, the Isetta didn't have a hood. The door across the front was all that separated the occupants from whatever came at them. The engine was under the seat. The Isetta was about as long as it was wide. It wasn't very wide.

The Isetta looked like something that had been flung from a carnival ride and took off down the road by itself. I guess back in those days you could get a license plate for just about anything. In all fairness I should point out, I am not describing the deluxe model. It was almost twice as big and even had a back seat of sorts.

WHAT WAS IT?

The driver of the Isetta stopped. He seemed to enjoy the attention. So do many inmates of mental institutions. The driver chatted with his vehicle's admirers. I got up close and even had the chance to sit in the marvelous machine. From the inside it was every bit as unimpressive as from the outside.

Imagine my surprise when nearly 40 years later I saw the reincarnation of this fabulous machine. Can just anybody buy one? Are they reserved for only those special people on television? Maybe it wasn't real. Maybe it was only a computer generated image. Those T.V. people do the darnedest things.

Maybe the one I thought I saw in 1957 wasn't real either. Maybe it's just a bad dream from my past – something I picked up while passing through the sixties. Lots of people had bad dreams in the sixties. Maybe I need a reality check. Does anyone else remember the Isetta? If you do, let me know. Please. Maybe we could get together and chat. We could explore the possibility of two people sharing the same hallucination.

If the Isetta was real, and if it is back – What does it mean? Have you seen any new Studebakers lately?

WHY PUNISHMENT?

Why punish criminals? Why must felons pay "their debts to society?" Is punishment what it's really all about? Who collects the "debt to society?" Who spends it? Punishment may satisfy a human desire for vengeance and retribution. Does it make the world a better place? Does punishment undo the suffering inflicted by the miscreant? Should it be our purpose to have our government inflict pain and suffering on wrongdoers?

If a criminal owes a "debt to society," How does time in prison, at great expense to that same society, repay the debt? Is it possible that all the rhetoric about punishment diverts attention from the real purpose of the so called "criminal justice system?"

Focusing on prevention of crime makes it is easier to see that action must be taken against the wrong-doers regardless of their mental conditions and life histories. We are not punishing them. We are protecting ourselves.

Retribution and vengeance are common reasons given for inflicting punishment. Punishment can also be used to alter a person's conduct. Punishment intended to correct or train is discipline. There is a big difference between retribution and training. It would be a start in the right direction to banish the word "punishment" from the vocabulary of the criminal law. Discipline criminals. Prevent crime. Don't be sidetracked onto the road of retribution and vengeance.

What harm comes from retribution and vengeance against criminals? It can make even the most vile criminal the object of sympathy. It's not the criminal's fault he committed murder. His father beat him. It's not the criminal's fault he mutilated. He is mentally ill. It isn't right to seek retribution and vengeance against these pathetic wrongdoers. We shrink from retribution and vengeance. We set the destroyer free to ruin the lives of more innocent victims.

WHY PUNISHMENT?

Does the state of mind of the perpetrator of a heinous deed really matter? Those who commit heinous deeds are mentally screwed up. Normal people don't do such things. Even those who commit less despicable crimes probably have a few loose connections in their mental circuits. Do mental defects make such people less of a threat to others? Not likely. If anything, their mental conditions make them more dangerous.

It may not be a dog's fault that it gets rabies. We hunt down the mad dog and dispose of it anyway. We don't kill mad dogs because we hate them, or because they are evil. We aren't punishing the dog. We do it because anything less would foolishly subject people to unnecessary risks of pain and death. Why should we be less concerned about protecting people from mentally sick humans who can be far more menacing than infected canines?

Focusing on prevention of crime makes it is easier to see that action must be taken against the wrongdoers regardless of their mental conditions and life histories. We are not punishing them. We are protecting ourselves. We can feel sympathy for them if we want to. We can be sorry their lives are so miserable. Still, if life in prison is what it takes to prevent more crimes – so be it. Unless we learn how to cure these messed up people, turning them loose should not be considered.

Let's forget about punishing criminals. Let's put terms like "punishment" and "debt to society" on the endangered species list. Let's make it clear that the real purpose for imposing sanctions on criminals is to prevent crime.

Persons subjected to imprisonment most likely feel punished. On the other hand, How do crime victims, their families, and friends feel? Should we subject the innocent to criminal depredations out of misguided sympathy for criminals? If someone must suffer, Is it not better that it be the one who committed the criminal act?

A Sporting Chance

Who doesn't want a sporting chance to win? Few are primarily concerned with how the game is played. We all want to be winners. Every game has its rules. Rules provide for orderly conduct of the contest. Sometimes rules are an attempt to place limits on the level of mayhem. All contests of humankind have their rules. The rules may not be written in books. Nevertheless, there are rules. Whether it's a football game, a battle for a parking space, or seeking the favors of a lady, there are rules. Rules to be followed. Rules to be broken.

Is a criminal defendant entitled to be tried under rules that give him a sporting chance to go free – no matter how obvious his guilt? No matter how serious his offense?

If the contestants are reasonably matched in ability, and have an even start, they will accept reasonable rules that bring some order to the fray. If one contestant knows that he has little or no chance of winning a fair, even contest, he will seek to alter the rules to improve his chances. Alterations take many forms. Who has not witnessed a game where a player suddenly claims the existence of some strange rule previously unknown to all? For some reason these newly discovered rules seem always to favor their discoverer.

Those who can't win with skill and ability, migrate to endeavors where the outcome is governed by pure chance. As an alternative they seek rules that take away the advantages that naturally go to those who have the determination and ability to win. One of the most obvious of these devices is handicap scoring, such as is common in golf and bowling. Can altering the scores really change the results? If the participants in the game are satisfied, Does it matter? Those who don't like the rules don't have to play. Everyone gets a sporting chance to win, regardless of ability and training.

This may be fine in fun and games between willing participants seeking a twenty-five-dollar trophy. Does the same hold true if the contestants are forced to participate? What if the contestants' property and lives are at stake? What if non participants will live or die based on the outcome of the contest? Is a criminal defendant entitled to be tried under rules that give him a sporting chance to go free – no matter how obvious his guilt? No matter how serious his offense?

What is the purpose of a trial? Is it a game where the prosecutor puts his wining streak on the line against a defendant exercising his right to a sporting chance to beat the rap? Is it too radical to suggest that a criminal trial should have something to do with seeking truth and justice? What justifies rules that interfere with the quest for truth and justice? In criminal proceedings are the real or imagined rights of the accused more important than the lives and safety of potential future victims? Does concealing vital, relevant evidence serve the cause of justice? Does it only provide a sporting chance for a guilty defendant? Or in some cases, may it protect the won-lost record of an over zealous prosecutor?

After reviewing reams of briefs, hearing the arguments of counsel, and considering the extensive research of its clerks, the U.S. Supreme Court rules 5 to 4 that the police erred. The evidence is inadmissible. To punish the police for their errant ways the defendant must go free to kill or maim again.

A cop on the street with his life on the line is supposed to instantly, correctly decide an issue that splits nine black robed justices sitting in the peace and security of their ivory tower in Washington. Is this justice, or is this lunacy? Are the rights of the guilty more sacred than the rights of the innocent victims?

Few would argue that a defendant, no matter how obvious his guilt, is not entitled to be treated fairly. What is fair? Considering what is at stake, Is it really unfair to trick or deceive a guilty defendant into telling the truth? If relevant evidence is obtained in a search later ruled improper, Is justice served by concealing the evidence from the jury? Is this fair to the jury? Is this fair to anyone? Where is it written that innocent men, women and children must be robbed, beaten and killed to atone for the real or imagined sins of the police? Only in the opinions of the courts, that's where!

Even the most ardent supporters of criminals' rights admit that the Supreme Court of the U.S. of A. made up the rule excluding valid evidence. This rule does not protect the rights of the innocent.

If no damaging evidence is found, What is there to exclude? If the police err, take appropriate action against the police. Don't punish innocent, future victims.

Sure, a court may find the "aggrieved" defendant unworthy of recovering any damages for the error of the police. The court may find that the police do not deserve to be punished. Does this less serve the ends of justice than setting free the guilty to again inflict their depredations upon the innocent? Is every guilty felon entitled to continue having a sporting chance to go free?

A GOOD SPORT

I watched the television news last night, though I don't make a habit of it. Most of what I see either bores me or makes me ill. This time illness was the order of the day. A Grand Rapids sports reporter drooled over the possibility of his city getting $25,000,000 from the State of Michigan to build a sports arena. How can even the dimmest bulb in Lansing so much as think out loud about the possibility of spending any tax dollars on something so frivolous? Why should you and I pay taxes to build a sports arena in Grand Rapids, or anyplace else?

I am sick of those poor mouthing, blackmailing sports teams. They pay tens of millions a year in salaries to players. Then the team threatens to leave town if we don't dig up the dough to build it a new place of business.

Adding insult to injury, the reporter mentioned that the $25,000,000 of taxpayer generosity to Grand Rapids could come as a spinoff from a Tiger Stadium fund for Detroit. I got the impression that the politicians were concerned they might be criticized for wasting money only in the Detroit area. Perhaps if they wasted a bit on the other side of the state the voters would be more forgiving. If balanced waste is what the voters seek, their bulbs don't burn real brighter either.

Wasting my money to build a new Tiger Stadium upsets me even more than wasting it on a minor league facility in Grand Rapids. It doesn't bother me more than wasting money on the Pontiac play pen of the Detroit Lions. That, of course, has been going on for years. Why am I so upset? I'm an unreasonable and greedy person, that's why. I am not pleased by the prospect of using my money to pay taxes to be used to pay the salaries of millionaire athletes. At least the athletes playing in Grand Rapids probably won't be millionaires.

Tax dollars used to pay for a stadium might as well go straight into the millionaire athlete's pockets. I am sick of those poor mouthing, blackmailing sports teams. They pay tens of millions a year in salaries to players. Then the team threatens to leave town if we don't dig up the dough to build it a new place of business. If teams cut salaries to a modest million or so, they could all build their own shops.

The athletes who are accustomed to ridiculous salaries wouldn't like the cuts. Are they going to quit? A few who are already rich might. The rest will fight just as hard as ever for the jobs. Even if the payrolls were cut to a quarter or less of present levels, the best players could still get rich. The rest would be making far more than they could anyplace else. Why should you and I work to subsidize a bunch of coddled millionaires?

Some will argue that we need these sports teams because they are good for the economy. That's garbage. It's an excuse, not a valid argument. Not valid to us taxpayers anyway.

All the stadium will do is draw consumers' dollars from other areas. When we tax an entire state to build a stadium, the very merchants who lose business to the stadium area merchants are taxed to build the stadium.

It has been said, a fool and his money are soon parted. This is no doubt true. The fact is, everybody and their money are soon parted. For some "soon" comes just a bit quicker than for others. The question isn't, if consumers will spend their money. The question is, Where?

A new Tiger Stadium or Grand Rapids stadium might benefit the surrounding merchants. If those merchants want to pay for the stadium, let them. That's fine with me. All the stadium will do is draw consumers' dollars from other areas. When we tax an entire state to build a stadium, the very merchants who lose business to the stadium area merchants are taxed to build the stadium. These victimized merchants should be even more upset than I am. At least the stadium won't cut my income.

All the stories you hear about spending tax dollars to generate business are a pack of lies. The tax dollars don't generate business, they just relocate it. For every winner there is a loser. As often as not, the losers have to pay to subsidize the winners. Let sports teams, and every other business, raise their own money and locate where it is most efficient for them to operate. Everybody will be better off, especially the taxpayers. A dollar spent in your community benefits the economy just as much as a dollar spent in a far off city.

P.S. I heard that the citizens of Los Angeles were faced with a bill of around $40,000,000 for repairs to their Coliseum. I have some advise for them. Let the teams that play in the Los Angeles Coliseum pay the bill to fix it up. $40,000,000 is less than the cost of signing one expensive athlete. How long should we cry if a couple of $40,000,000 athletes have to take 50 percent pay cuts to pay for repairs to the shop where they "earn" their millions? How would the poor fellows ever make it on a mere $20,000,000? Could you survive on such a pittance?

MUST WE CONCEDE TO TERRORISTS?

A recent news item reported that Kennesaw State College in Marietta, Georgia renamed its mascot. Why? Who cares? Most of us never heard of Kennesaw College. Does it matter that "Hooter" the owl was branded as politically incorrect?

Language is ever changing. Many words have more than one meaning. Look in your dictionary. See how many words you can find with only a single definition? In my dictionary the word "single" has seven definitions, two of which contain sub definitions. It might be nice if each word had only one meaning. Then again, it might not. Do we need tens of thousands of new word to put on spelling tests?

Will we continue to appease the verbal terrorists, or will we stand and fight? If we lack the courage to rise to the cause, Will we someday stand at a verbal Dunkirk with our backs to the wordless sea? Will we be doomed to watch helplessly as the verbal Gestapo strip the last shreds of our vocabulary from us?

When a word is first used in a new way, we call it slang. English teachers discourage the use of slang. Most slang passes. Some gains reluctant recognition by those people who make dictionaries. The dictionary usually labels the new usage as slang. Certain words bear the even more infamous designation of "vulgar slang."

Does the dictionary drop all of the old meanings of words branded as vulgar slang? Not by a long shot. The *American Heritage Dictionary* lists 18 possible usages for a word that is familiar to all of us. Two of those usages bear the infamous brand of "vulgar slang." Are we going to quit using this word for the other 16 purposes? If we do, What will we call the act of a pin puncturing a balloon?

Should we all be forced to give up the common usage of a word just because it is held prisoner by verbal terrorists in the dungeon of vulgar slang? To do so is to appease those who degrade our language. Even the location of this nether land of vulgar slang is not clear. My dictionary recognizes four usages of the word "vulgar." They range from "language spoken by the common people" to "obscene or indecent." Perhaps three of these usages should be eliminated. Which three?

What will happen to our language if we abandon to their captors all the words held hostage? Consider some of the words already imprisoned on that literary Barbary Coast? Far more than owls risk losing their names.

What are we to call that metal fastener with threads that we twist into a piece of wood? Must half the voice of the bell be muted? Must the bell now only go "ding?" For that matter must the famous, or infamous, pastry named for the sound of the bell have its name truncated? Getting back to the problem of the owl. Who will be responsible for editing the offending town name out of all those episodes of "Petticoat Junction?"

More than words are at stake. Must we give up the sixteenth letter of our alphabet? To shorten the count I'll give a clue. The fifteenth letter of the alphabet is "O." For those who are numerically or alphabetically challenged, the clue may not help.

There is nothing new about abandoning words to rot away in captivity. In the days of my youth chickens had white meat and dark meat. "Breasts" and "thighs" still languished in that literary dungeon. Those prisoners escaped. The dungeon keepers now try to compensate for the loss by capturing poor "Hooter" the owl.

Will we continue to appease the verbal terrorists, or will we stand and fight? If we lack the courage to rise to the cause, Will we someday stand at a verbal Dunkirk with our backs to the wordless sea? Will we be doomed to watch helplessly as the verbal Gestapo strip the last shreds of our vocabulary from us?

There is far more at stake than the loss of words. The real threat from the verbal terrorists is not that they take words from us. Words are replaceable. Verbal terrorists march in lock step with the "thought police." They are mostly the same people. Verbal terrorist is just their night job. They try to control our thoughts, not just our words.

Certainly there are some thoughts the world could do without. There is nothing wrong with trying to get another person to think differently. We all try from time to time. The danger comes when a group resorts to force and intimidation because it is just too much trouble to convince others to change. Ideas that can be spread only thorough force and intimidation are unworthy of being spread at all. If the ideas championed by the verbal terrorists and thought police cannot stand on their own, they shouldn't stand at all. If the ideas have merit their proponents do a disservice to their own causes by associating them with the worthless ideas that can be spread only through intimidation. We tread a dangerous road when we tolerate the control of thought by intimidation.

THE EXTENSION LADDER

My two-section extension ladder is great for cleaning eaves' troughs and painting upstairs windows. I use the bottom half for those lower jobs, like getting up to the garage roof. It's a bit difficult to use the top half by itself. I could dangle it from the peak of the house. I could still climb the ladder to the eaves, if I could get past that missing bottom half. Standing on the ground jumping up and down probably wouldn't do it. If I were that tall I'd be a professional basketball player.

Why should you care about my extension ladder? No reason I can think of. My extension ladder reminds me of something about which you might have at least a whit of concern. The topic is on all our minds. Have you heard anyone mention education lately? What do extension ladders have to do with education?

Many parents believe they have fulfilled their educational responsibilities by pushing the kids out the door toward the school bus. Many teachers like it that way.

Education comes in two sections, the parents' part and the teachers' part. Parents are teachers too. They just don't have the title. Parents are the bottom half of that educational ladder, while teachers are the top half. If half of my ladder lays on the ground and half leans against the house, I'm not going to get to the top of anything. Likewise, the educational ladder doesn't work when the bottom half is missing or broken. To reach the top of the house of education requires the entire ladder.

Many parents and teachers have lost sight of the importance of the bottom half of the educational ladder. We live in the age of the specialist, the professional. Television breaks, call the repairman. Muffler falls off, head for the garage. Bugs in the house, call Orkin. Kids need teaching, send them to the teacher. Many parents believe they have fulfilled their educational responsibilities by pushing the kids out the door toward the school bus. Many teachers like it that way.

If we are going to fix education we must repair both halves of the ladder, and get them back together. In education, just like with my ladder, the bottom half is the most important. Without it the top half doesn't count for much. Education reform, like charity, begins at home.

Heavy-handed tactics and arrogance of school leaders have alienated many parents from the educational process. Schools that don't support the values and aspirations of the parents cannot gain the support of the parents. Using schools for social engineering is worse than useless. The social engineering doesn't work. Education suffers.

Even if we have two fine halves of a ladder, we still aren't going to get to the top unless we get the two parts together. Parents and teachers must each recognize, and accept, their own part of the task. Then they must get together to see to it that the children climb up the ladder.

Sure, there are alternatives. I might be able to climb out an upstairs window and make it to the top of the house without the bottom half of the ladder. I might fall and break a leg. I might be able to climb up a drain pipe if the top half of the ladder was broken or missing. I'm not that eager to climb. Many children aren't that eager to get an education. Some may make it in spite of part of the ladder being missing or broken. Many are not that talented or motivated.

The public schools are less than perfect. A little common sense and discipline would fix most of the problems. The performance of the top half of the ladder isn't the biggest problem. It isn't nearly as broken as some believe. For the children who have a good bottom half to their ladder, the top half is adequate, even though not excellent. Prodding from concerned parents often improves the performance of the teachers. Tinkering with the top half isn't going to do much for the children who have a ladder with a bottom half that doesn't function.

The biggest problem with the top half of the ladder is that schools aren't doing enough to bring the parents back into education. Perhaps schools shouldn't have to do this. On the other hand, the schools aren't blameless. Heavy-handed tactics and arrogance of school leaders have alienated many parents from the educational process. Schools that don't support the values and aspirations of the parents cannot gain the support of the parents. Using schools for social engineering is worse than useless. The social engineering doesn't work. Education suffers.

To improve education, we must find out why so many parents drop out of the education of their children. We must then bring the parents back. Both the parents and the schools should be concerned about the entire ladder, not just half of it. We must never forget that parents are more important to education than school teachers. Parents must provide the foundation for education. Without a good foundation successful building is impossible. Many parents educate their children at home without professional teachers. It works. Teachers usually are unable to provide a good education when the parents don't do their part.

Public schools are losing the parents in two radically different ways. Some parents send their kids to school but make no contribution to their education. Other parents abandon the public schools for private ones or home schooling. In the latter cases the children don't suffer. When the parents do nothing, the children are the big losers. The dilution of the interest in learning affects all the children in the school.

The biggest problem with so-called education reform is the big push for more state and federal standards and control. The main effect of state involvement is to push the two halves of the ladder further apart. Education would work just fine without the state and federal governments being involved at all. After all, Aren't some of our best schools those of the private variety? What advantage do private schools have over public ones? The biggest advantage is that the parents who put forth the effort to send their children to private schools are concerned enough to provide their part of the ladder. Without the parents, all you have is the top half of a ladder swinging in the breeze.

A New Era

Why should school teachers invest their effort to bring reluctant parents back into the educational process? How will this benefit the teachers? Survival of the teaching profession as we know it depends on healing the rift between parents and teachers. With today's technology it will soon be possible for just about all families to educate their children at home with minimal professional assistance. Many do not seem to realize that the main function of the teacher is providing motivation and discipline. This is the area where public schools are the weakest. Parents can easily do a better job. Any assistance the parents need with subject matter will be available through computers.

Public schools enjoyed a brief period of minimal competition. Any institution not held accountable by competition, crumbles within.

I was part of the last generation to share the educational experience of the one room country school. If schools don't clean up their act, the present generation may be the last to ride the school bus. That wouldn't be all bad. Interactive television networks already exist and are in use at some schools. Expand their scope and extend these networks to the home, then schools will have to produce or die. Unless the traditional schools can provide a benefit to justify their cost and inconvenience, people will simply refuse to support them.

The nay sayers from the ranks of the teachers' unions and the social engineers cannot stop choices in education. Private schools have grown for decades. The interactive computer-television option will be born, no matter how desperate the attempts to abort it. Competition is coming to education. The question is not, If? The question is only, When?

Every family will not personally monitor the education of their own children. Parents may take turns monitoring the education of the children from several families. A group of families may hire someone to supervise all the children. The monitor will not have to provide the teaching expertise. Thus, children of all ages and levels of achievement can be in the same group. All can proceed at their own paces. The results may look a lot like the rebirth of one room, neighborhood schools. Most children won't leave the neighborhood or the apartment complex for their education. Some children with special problems will still be sent to the teachers.

Sports and other group activities will be separate from school. There's nothing new about this. We already have Little League and other sports programs outside the schools. Scouts and 4-H are major programs. Such activities will expand to fill the gap left by the demise of the central school.

The new schools may be either public or private. At first they are almost certain to be private. The present public schools can play an important role in the new system. Strong, positive involvement from the public schools could ease the transition to the new era.

Unless public education takes bold, strong action to become user-friendly and sell itself to the parents, it will wither and die. The public schools will become depositories for the children of parents who don't give a hoot about education. Public school teachers will be nothing more than wardens filling the gap between the home and welfare or prison.

Public schools enjoyed a brief period of minimal competition. Any institution not held accountable by competition, crumbles within. Look at what happened to giant companies like General Motors and IBM during their fleeting periods of market dominance that relieved them of the need to compete. When the competition returned, these companies were ill prepared to handle it. They endured years of agony before beginning to learn to compete again.

Public education now stands at the same cross road. Learn to compete, or perish. In competitive markets the customer is always right, whether we like it or not. Public education must get its house in order and woo the customers in order to survive. It must redesign its product. Never mind that the current product isn't so bad as some people think. The point is that better, more user friendly, products are coming to the educational market. Studebaker wasn't a bad car. Still, it is no more.

If public schools are up to the challenge, they can be an important part of education for generations to come. American auto manufacturers lost their market dominance during the seventies and eighties. Public schools are destined for the same fate. American auto makers learned to compete and now hold a solid market share. Public schools can also hold a solid market share. To do so they must earn it. They are not going to be the only game in town. To survive they must convince parents that public schools have the best game in town.

Most attempts to "improve" public schools today are the equivalent of rearranging deck chairs on the Titanic. I fear that most people in public education do not even know they have hit an iceberg. The crews of the public schools believe they are indispensable, therefore indestructible. The longer the public schools fight back the tide of change, the more devastating the end will be when the tide finally has its way. Beating one's head against the stonewall of reality damages the head far more than the wall. There are going to be radical changes in the way we educate our children, with or without the public schools.

We will all be better off if the public schools change and prosper. They can add an important mix to the educational market. To do so they must learn to compete, rather than trying to eliminate the competition. Suppression of competition only works for a while. If you don't believe it, ask AT&T.

We cannot perform an educational transplant that will restore health to the public schools. The schools must cure themselves from within. The administrators and teachers will not undertake the painful cure without motivation. Competition is the only way to provide that motivation. Competition will threaten the survival of public schools that offer an inferior and expensive product. Nothing motivates like a threat to survival. When teachers unions see that improved public education is the only way they can survive – they will give up trying to eliminate competition and join in the struggle to improve education. (Just before press time, the Michigan Education Association announced it was joining the charter school venture. Previously the MEA boycotted a college and its graduates because the college supported charter schools. The sincerity of the MEA is yet to be proven. If it follows through, this could be a major breakthrough for quality education.)

BAN ASSAULT RIFLES?

A congressman from Wisconsin drew national attention and a bit of ridicule for pointing out that more people are killed each year with hands and feet than with assault rifles. I don't have the numbers, yet I have no doubt the congressman is correct. It would surprise me to find that the number of people strangled and kicked to death does not exceed the small number killed with this thing called an "assault rifle."

I would not be surprised to find that more people are killed with hands and feet than with all rifles. Rifles aren't very high on the criminal's list of favorite weapons. Don't forget, so far no one has managed to write a definition of an assault rifle that isn't either full of holes, or else includes many rifles traditionally used for hunting and target shooting. That's why legislators engage in the rather useless acts of banning assault rifles by name.

The questions we should consider are: Will banning assault rifles do any good? Should we ban assault rifles? If the answer to the first question is no, the answer to the second question should be obvious. We should not encourage government to enact useless laws.

You may have heard about the California teenager who legally changed his name to Fishing In America. Gun manufacturers are no less ingenious at name changing than was the new Mr. America.

One tactic of those seeking to ban assault rifles is having victims and their families recount before Congress and TV how horrible it is being shot with an assault rifle. This tactic stirs emotions. It adds nothing to intelligent consideration of the problem. Anyone who needs convincing that shooting people with assault rifles is a horrible thing needs far more than a lecture.

Many assault rifle incidents involve only one or two shots. It is difficult to understand why a person shot with a single round from an "assault rifle" is worse off than if he were shot with some other kind of weapon that could just as easily deliver the same bullet.

The questions we should consider are: Will banning assault rifles do any good? Should we ban assault rifles? If the answer to the first question is no, the answer to the second question should be obvious. We should not encourage government to enact useless laws. We get enough useless laws without encouraging them.

There are lots of horrors in the world. Occasionally a driver deliberately crashes his vehicle into a crowd killing and maiming. Even more often drunk drivers kill and maim. Should we ask the victims' families to urge us to ban automobiles and alcohol? You may recall, we banned alcohol once. The ban wasn't a rip roaring success.

Millions of people own and lawfully use these things called assault rifles. An extremely small percentage of these rifles are ever used to commit crimes. I hope there are none among us so naive as to believe that any crime will go uncommitted for want of an assault rifle. If assault rifles aren't available, the criminal will use some other weapon.

That other weapon might do less harm. In some instances the substitute weapon could do more harm. The assault rifle isn't the ultimate weapon. Many crimes committed with assault rifles could have been even more deadly if the perpetrator had used other readily available weapons. If I were teaching Mayhem 101, I would elaborate. I see no point in giving advice to would be criminals.

I would support a ban of assault rifles if there was a reasonable possibility that it would take a bite out of crime. Taking these guns away from millions of law-abiding citizens would be a major intrusion in their lives. If we could prevent a few hundred murders a year, we could afford the price. I would just as quickly support the ban of alcohol if I thought the ban would prevent the tens of thousands of deaths caused each year by alcohol.

Wanting to put an end to something and being able to, are two very different things. We gain nothing by embarking on emotional campaigns that lead only to failure. People are the problem. We can solve our problems, only by dealing with the problem people. Trying to hide the tools of crime won't work. If criminals want assault rifles, they will get them.

Certain people like to tell us how much they feel and how much they care. There are millions of gullible voters who gush and rave about "feeling" and "caring" politicians. Feeling and caring, when not tempered by disciplined thought and reason, are loose cannons firing at us all. It is part of the "pure heart, empty mind" syndrome. Is it better to be governed by caring people who don't think, than by thinking people who don't care? Ideally we would find leaders that both care and think. It's easier to fake caring than it is to fake thinking. Might I suggest that we all keep this in mind when we vote. If we don't see through the scams of the political con artists, we will again end up being governed by the best con artists.

INDIGNATION

A seventeen year old in New York State was accused of sneaking into a park and beheading a swan. According to the *Associated Press* report:

> "[The accused] has received threatening calls and letters since the swan was killed. . . . Nearly 100 people, the majority angry, showed up for the boy's arraignment Thursday night.
>
> About 30 people peppered the boy and his lawyer with verbal abuse as they left the courthouse."

Some people believe this is a lot of fuss over a bird. Maybe it is. If the teenager had burned down the park maintenance building, Would there have been a public display of anger? For that matter, if he had killed or raped, Would anyone other than family members have shown up for the hearing?

If we want to prevent crime, we must change our attitudes, not our laws. If we again treat criminals and their associates with contempt and derision, crime will wither.

This incident does more to explain the crime problem in America than a whole stack of books could. We don't abhor crime and criminals anymore. I'm not suggesting we should show more sympathy and understanding to swan killers. We should show far less to other criminals. Criminals, and often their families, used to be objects of scorn and contempt. This social pressure did far more to control crime than did laws or police. Crime flourishes, not for lack of laws or lack of police, rather because many find crime "cool" instead of contemptible.

Who could have missed the ruckus about the caning of a teenage vandal in Singapore? The low crime rate in Singapore isn't because of caning. The public attitude toward crime in Singapore makes serious punishment acceptable, even for what we Americans often call minor offenses. It's this attitude that prevents crime far more than the caning. The main value of caning is, it reinforces the attitude.

If we want to prevent crime, we must change our attitudes, not our laws. If we again treat criminals and their associates with contempt and derision, crime will wither. The generation raised in the anti criminal environment will have far more respect for people and their property.

We cannot afford to reserve our contempt for armed robbers and murders. By the time the criminal reaches the big leagues, it's too late. Youthful offenders who engage in petty shoplifting, vandalism, and the like, must be visited with an outpouring of scorn and contempt. Treating these offenses as overlookable, youthful indiscretions is a mistake for which we pay and pay and pay. Sugar coating youthful crime by calling it juvenile delinquency is a major mistake. Criminals are criminals and should be treated as such, whatever their age. The biggest cause of crime is the failure of parents to raise children with respect for the lives and property of others. Scorn, contempt and ridicule are perhaps the only effective ways to motivate neglectful parents to instill law-abiding attitudes in their children. Making the lawless children social outcasts will put direct pressure on them too.

If all youthful offenders who commit "minor" offenses were visited with the same contempt as the accused swan killer, crime in America wouldn't be any more of a problem than it is in Singapore. For the social pressure of contempt for crime to work it must spring from the entire community, not just a few people. We have a big job ahead of us. Government can't do that job. Either citizens by the millions will do the job, or it will go undone. If the job goes undone, all law-abiding citizens will soon be living in fortresses.

GRAB BAG

In 1845 Congress appropriated $45,000 to ship camels to the Southwest. Stung by the failure of this program, Congress has since dedicated itself to shipping jackasses to Washington.

* * * * * *

It's almost that season again. Time for the voters to renew that eternal question to those congresspeople sent to Washington. "Why don't you bring me flowers anymore?" It's the wrong question. We should ask, "Why don't you get out of our way and let us grow our own gardens?"

Flowers grown in Washington wither and die long before they reach us. This is most fortunate for those politicians. Many voters are unable to recognize that the withered "flowers" were never anything but weeds.

Would we rehire the florists if we knew that they grew only weeds? If we want flowers, we must grow them at home. Even real flowers indigenous to Washington would not flourish in our local soil.

If "Fish" can be a last name, Does it not have the right to aspire to be a middle name? So what if it adds an "ing." Achievement requires sacrifice.

The politicians in Washington can help. They can send us fertilizer. That is the one thing they do really well. If we expect more, we are doomed to disappointment. Don't forget. We are paying the florist's bill although we are getting only dead weeds.

* * * * * *

A young man in California legally changed his name to Trout Fishing In America. Supposedly he did this on his own. Considering the volume of monologue material the new name provided for Jay Leno, one could speculate that he put the young man up to it.

Most people seem to find Mr. America's new name a bit strange. Is it? Trout as a name is not unheard of. I recall a Detroit Tiger pitcher named Paul "Dizzy" Trout. Girls sometimes have "America" as a first name. If a young lady named America married into the Trout family, Wouldn't her name be America Trout?

If a woman can honestly come by the name America Trout, What is wrong with a man being named Trout America? There should be at least a little difference between men's and women's names. If the feminists and the Lorena Bobbitts of the world have their way, names may be all that's left to distinguish between men and women.

What about those middle names? Most people don't have two middle names. Yet, some people do. How many of them are "Fishing?" "Fish" as a last name is not uncommon. My phone directory has more Fish than I catch on several fishing trips. If "Fish" can be a last name, Does it not have the right to aspire to be a middle name? So what if it adds an "ing." Achievement requires sacrifice.

I admit that "In" doesn't have much of a history as a name. Two letter names just haven't made it. My unscientific survey shows that less than 0.02 percent of all names in this country are of the two letter variety. Most two letter names in this country belong to the Chinese. Might I suggest that the absence of two letter names shows a patten of discrimination based on anti Chinese sentiments.

If the federal government catches on, we face name quotas to rectify this inherent discrimination against the Chinese and two letter names. This is not a joking matter. In some countries parents must select their child's name from a list approved by the government. Do you want the feds telling you how to name your children? Mr. America took one small step toward rectifying the problem before the feds impose their brand of a solution on us. On behalf of us all – Thank you, Mr. America.

AL IN WONDERLAND

This report is dedicated to all those outstanding financial institutions that serve us so faithfully without error or mistake. It is also dedicated to all who in a moment of weakness may have thought one of these sophisticated and flawless establishments did err.

One of my sons borrowed my credit card. Then he discovered his billfold was missing. After I reported the card as lost, the bank dutifully canceled the card and set about to replace it. One selling point for this card was, there would be no charge for replacement of lost cards. Being the unreasonable person I am, it annoyed me when I received a letter advising me that I was being charged $10 for new cards.

If every time I dial 1-800 I get $10, I might be on to something. I had only called once a month. Could I call every week, or even every day?

I was so annoyed that I dialed the 1-800 number to express my displeasure while canceling the card. When the monthly statement arrived, there was the $10 charge big as life. Directly below it was a $10 credit canceling the charge. It seems the bank had corrected its error even before I called. All is well that ends well, so I've heard.

The next month I received a statement showing a $10 balance. No, they weren't still trying to stick me for the $10. The bank now claimed to owe me $10. It had again deducted the $10 charge. They didn't say how they proposed to pay it, or how much interest I would get. I ignored the statement, fanaticizing that by next month the balance would float back to zero. It didn't.

Having told the bank to close the account, I destroyed the new cards when they arrived. Thus, it would be a little difficult for me to spend the ten dollars. Facing the prospect of getting repeats of that monthly statement forever after, I again dialed 1-800. I told the young lady who answered the phone that I had a problem with my account. I could feel her brace herself for the onslaught. As I launched into my story I sensed, all the way from Delaware, that she was not used to having customers call to complain that their bills were too low. She immediately spotted the double $10 credit and assured me it would be taken care of so I wouldn't get any more statements.

Again at the first of the month the familiar envelope arrived. Perhaps they just had to send me one more statement to show that my balance was now zero. If life were only so simple. They did make a $10 adjustment to my account, just like the young lady promised. The only problem was, now they "owed" me $20.

If every time I dial 1-800 I get $10, I might be on to something. I had only called once a month. Could I call every week, or even every day? I called and asked. The woman who answered the phone failed to detect any humor in my inquiry. I had the uneasy feeling that she didn't even understand the question. She could see that I had a $20 credit balance. She didn't seem to understand how I got it or much care. She did want to help with my unusual problem. She could only think of two ways to zero out my balance. I could spend $20 (I hadn't told her I cut up the cards), or I could request a check for $20. There was no other way. I had the feeling that either route would lead down a twisted trail I had no desire to tread. Just getting one more piece of junk mail each month was looking better and better.

I tried to concede and hang up. The clerk was tenacious as a pit bull, though much more pleasant. She kept urging that I request a check to balance out the account. She even pleaded. She said she was at my mercy. She was trying so hard I didn't want to hang up on her. I finally told her to go ahead and send a check if that was what she wanted to do. Why do I think I'll end up regretting my choice? I can hardly wait to see what the mail brings next month. If you hear about the national debt being paid off, don't even ask whose credit card it was charged to.

Now I know why banks never make mistakes. They can't afford to until they discover how to correct them. You can imagine where this might be headed if the bank had made a serious mistake.

Banks, even big ones, are much smaller than the federal government. Banks are also motivated by competition and the need to make a profit. Not only that, simply posting debits and credits to an account is not exactly a matter of mind boggling complexity.

When a bank has this much trouble managing a credit card account, imaging the fun you could have dealing with nameless, faceless computer jockeys in a federal health care bureaucracy. When the bank screws up I can switch to one of the other banks that solicit my business on about a weekly basis. Federal health care wouldn't just be the only game in town, it would be the only game in the country. Federal health care probably would cut costs. Everyone would die before any services were provided.

HOW GREAT ARE WE?

America is a mighty and powerful nation. No one questions this. Some among us suggest, America's greatest days are already history. They worry that, like all great civilizations of the past, America has begun to decline.

Today no one questions that America is the world's mightiest military power. There is more concern about our economy. The economies of many nations are growing. Those growing economies provide competition for the U.S. of A. In some cases we are not handling that competition real well.

The economic strength of other nations is gaining on us. A careful look shows that gaining is exactly what those nations are doing. Technology, productivity, output, whatever you want to measure – those other nations are still trying to catch up. The footsteps we hear are still behind us.

The United States is still the world's foremost economic power. We are being challenged. We will continue to be challenged. Instead of fretting about the competition, we must get in shape for the race. We will be better off in the long run because of the competition.

What is really troubling us is, we haven't gotten used to hearing footsteps. At the end of World War II the U. S. was so far ahead of everyone that we were without competition. America was a world class sprinter running against a bunch of kindergartners.

The kindergartners grew up. The sprinter is still champ. He looks over his shoulder and worries about tomorrow's race. He should worry. Even a world class sprinter running for years without competition will not maintain top condition. He will lose his ability to handle a real challenge. Any star athlete knows that the quality of the competition determines the level of his own performance. If the competition is weak, his own performance will not be inspired.

Loss of competition is just as deadly for business. Companies seek to become monopolies. Yet, if they succeed, their success will ultimately be their undoing. Without competition the performance of any business slips. It loses the competitive edge that brought the initial success. Look what happened to General Motors and IBM after a few years of dominance in their worlds? If their dominance had continued for a few more years, those giant companies might well have been weakened too much to survive.

The entire economy of the United States suffered through a lack of serious outside competition in the post World War II period. Sure, there was the Volkswagen and a few things like that. They were no more than mosquitoes buzzing an elephant. When the competition returned in a serious way during the seventies and eighties, we as a nation had to relearn the game. We are starting to catch on.

We should not resent or fear the competition. We should be thankful that it came when it did. The longer we avoided the competition, the less able we would have been to respond to it.

The United States is still the world's foremost economic power. We are being challenged. We will continue to be challenged. Instead of fretting about the competition, we must get in shape for the race. We will be better off in the long run because of the competition.

Getting in shape takes effort. It takes discipline. There is a price to pay. We must be less indulgent. We must be lean and tough. There will be sweat and sore muscles. Non productive jobs must be eliminated. Like it or not, the low skill, high pay jobs of the post War era have no place on the body of a well conditioned economy. We must go back to the era of more skill, more pay. New jobs will replace the ones lost.

During the years of economic flab I often heard the comment, "Why should I get more education? I can make big bucks right out of high school without wasting more time in school." Those days are over. Those who want higher paying jobs must offer greater skills in return. Those skills may be learned outside the classroom. They must be learned somewhere.

A high school education is now more essential than ever. It is not a passport to success. By itself a high school education is the ticket of admission to the theater of minimum wage. That is the first big step above unemployment. Note that I said "high school education," not "high school diploma." A diploma without the education is a ticket to nowhere. After graduation it is up to the graduate to take the next step. As in any race, the participants must run just to hold their own.

America isn't yet declining as a world economic power. The other nations are catching up. If we cut the flab and get to work, decline is not inevitable. If we don't rise to the challenge, the world will pass us by. Our new national slogan of, "What can I get?" must be replaced with the old one, "What can I do?" If we again become a nation of doers rather than getters, we will do just fine. We will do our share of "getting" too.

THE LIMIT OF THE LAW

Murder, robbery, rape, vandalism, child abuse, illegitimacy, welfare, and innumerable other social disorders are on the rise across our nation. The "good old days" weren't so good as we may paint them. Still, it is inescapable that irresponsible conduct is spreading over our society like lava from an erupting volcano. Why? Some say, poverty causes all our ills. A recent article in *U.S. News* reported on a study that concluded 80 percent of the population lived below the poverty line in the 1890s. Be this conclusion right or wrong, there can be no doubt that most of the "poor" today have access to far more resources than did the average citizen a century ago. If poverty causes crime and irresponsibility, the eighteen hundreds would have been anarchy.

For the most part, people conform their conduct to what's expected by their peers. When those peers don't expect and demand responsible conduct, passing laws will not correct individual behavior.

Seeking solutions to our problems, we pass more laws. If state laws don't work, pass federal laws. If federal laws don't work, pass more of them. If passing laws brought about responsible conduct, we would be living in the most responsible society that ever existed.

Law enforcement never has, and never will, create a law-abiding society. For the most part, people conform their conduct to what's expected by their peers. When those peers don't expect and demand responsible conduct, passing laws will not correct individual behavior. When most people support and demand a certain standard of conduct, few people risk the social censure that goes with violating the will of the community. Laws reinforce the community standard and can be somewhat effective to bring a small minority in line with that standard. Laws are totally ineffective when a major segment of the community does not support them. Law is the truck that follows behind to pick up the stragglers along the march of civilization. If there are very many stragglers, the truck simply isn't up to the job.

One of our first major encounters with this inevitable principle was prohibition.

Even dictators recognize the weakness of laws. If laws really worked, dictators wouldn't have to resort to information control and brain washing. Despotism succeeds, not because laws control the bodies of the citizens, rather because propaganda and censorship control the minds. The population may be deceived, yet most willingly follow the despot in charge. To believe that laws can be more effective in a democracy is to deny reality.

The only hope for bringing order and responsibility to our communities and nation is to unite the people in a common purpose. Violence, drunk driving, child abuse, vandalism, or any other evil, will be controlled only when the overwhelming majority of our citizens stand openly against it. Saying we are against it is not enough. Those who are truly against something don't look the other way when friends and neighbors do it. We think less of those who violate our principles, and we let them know.

For so long as society provides safe havens for the irresponsible, irresponsibility will flourish. Understanding and sympathy for those who engage in irresponsible conduct, encourages and supports that conduct. Until we demand accountability at the personal, individual level, it will serve little purpose to demand it with laws. We must again treat the irresponsible as social lepers.

First we must reach a consensus as to what is irresponsible. Until we reach that consensus, and live our lives accordingly, our civilization will continue to deteriorate. There is room for diversity, but not when it comes to the basic principles of civilization.

We should attempt to rehabilitate those whom we can. The beginning of rehabilitation is the recognition by the irresponsible individual that he is at fault and must change. If we continue to be apologists for irresponsible conduct, we are doomed.

WE GET LETTERS

I wonder how many were blessed with a copy of the letter I received awhile back. It began, "Kiss someone you love when you get this letter and make magic. This letter has been sent to you for good luck." The letter bore a Lansing, Michigan postmark and explained how I must send out 20 copies in 96 hours, or else be cursed. What happened to the good old days when chain letters only required the dispatching of seven copies? I guess inflation strikes everywhere. The writer detailed how death and pestilence befall those who ignore the letter. He further explained how good fortune is the lot of those who dutifully pay $6.40 for postage to send the letter on. (The letter didn't mention the $6.40.)

The most interesting part of the letter was its explanation of how it was written by Samuel Anthony Pearce, a missionary from South America, and that the original letter is in New England. When Mr. Pearce wrote the letter, How did he know about all the good fortune and misfortune that would visit people who would eventually receive the letter? Maybe he called the psychic hot line.

What if I mailed out 20 copies and every recipient down the line did likewise? The eighth generation mailing would contain 25.6 billion copies. That would be four letters for each man, woman and child on earth, with enough left over for most of the dead people who vote in Chicago.

I ignored the letter though I kept it in my file. I'm still alive to write about it. On the other hand, I haven't won the lottery, as some recipients of the letter supposedly did. Of course, I haven't bought any lottery tickets either. If you check the odds on Lotto you will discover that those who buy tickets have little more chance of winning than do those who don't buy.

Why, after a couple of months, am I inspired to write about a rather dull and useless letter? Today my wife received a copy of a very similar letter postmarked at Gaylord, Michigan. Comparing the two letters piqued my interest. They clearly have a common origin.

They describe the same good and bad luck. They are both a bit unclear about whether the man whose wife died after she won the lottery was having good or bad luck.

The alleged author, Samuel Anthony Pearce, a missionary from South America, was now Saul Anthony Decroe of South Venezuela. Most other names, along with many other words, were slightly altered. One tragedy was omitted. My letter tells of a person who received the letter in 1988 and then won the lottery. My wife's copy says the letter was received in 1958. If I recall correctly, about the only lottery in 1958 was the Irish Sweepstakes.

Now folks, it's quiz time. Does sending altered copies count, or do these stumbling scribes still incur the wrath of the letter gods? Is it possible that they only have their luck altered in a manner commensurate with the errors in the copies they made? The even bigger question is, if my letter isn't a true copy of the original, Does it count at all? Could it be that even if I sent out the 20 copies, I would be getting my hopes up for nothing? How can I sleep at night, or in the daytime, knowing there may be counterfeit chain letters in circulation? Or could it be that if the errors are made in good faith, they will be forgiven?

What if I mailed out 20 copies and every recipient down the line did likewise? The eighth generation mailing would contain 25.6 billion copies. That would be four letters for each man, woman and child on earth, with enough left over for most of the dead people who vote in Chicago. The postage for that mailing would be over $8.1 billion. If there was one more mailing, the postage would be over $162 billion. At one mailing a week this would only take nine weeks. I think Bill Clinton started these letters. What better way to eliminate the federal deficit by election day?

WHERE ARE WE HEADED?

Fifty years ago I walked down the dusty gravel road out of summer and into the one room school nestled among the maple trees. My formal education began as it had for several generations before me. Change is not in the vocabulary of a six year old. The schoolhouse had always been there and always would be. Actually the schoolhouse is still there, though the bell long ago ceased to toll. What was once the Mecca of education for the community is now just another residence.

Though the handwriting was already on the wall, I knew not that by the time I completed my formal education, small neighborhood schools would go the way of the dinosaur. The changes in the form of education over the past 50 years overwhelm the changes during any comparable period in history. Unfortunately, many of the changes were not for the better. Education, like everything else, is the product of the environment in which we live. With the radical changes in that environment, drastic changes in education were inevitable.

New products are bursting over the horizon of the educational market. Those who live in the halls of traditional education seek to repel the newcomers to sustain the education that has served us reasonably well. Change cannot be held back when the market demands it. Education will change dramatically over the next 50 years.

The changes in our social and technological environment continue unabated. It is inevitable that these changes will carry over into education. When those bright eyed kindergartners who so recently completed their first journey on the yellow bus look back on education 50 years from now, What will they see? No one can give a complete answer to that question. One safe bet is that in 2046 the school bus will be as rare as the one room schoolhouse is today.

A news report told of the closing of four, single room schools in Vermont. It is indeed ironic that these schools, that many see as dinosaurs, survived until most of the reasons for their demise had passed, until the time the benefits of smaller schools are growing more and more apparent. Most of the problems of education today spring from the schools being too distant (Not physically distant, rather emotionally distant.) and from concentrating too many students in one place.

Over the years we defended the distance and crowding as essential to bringing together enough students to support the variety of courses essential for a complete education. Loss of security in the schools, and alienation of the parents and children served, were just part of the price we had to pay for a school system big enough to meet the needs of the students. People are growing less willing to pay the ever increasing price. When a necessity of life comes in only one brand, we complain and pay the price, no matter how high.

New products are bursting over the horizon of the educational market. Those who live in the halls of traditional education seek to repel the newcomers to sustain the education that has served us reasonably well. Change cannot be held back when the market demands it. Education will change dramatically over the next 50 years. Even 25 years from now the educational system will little resemble what we see today.

The question is not, "Will there be change?" The question is, "What will the changes be, and who will shape them?" At a time when balkanization of our society is one of our greatest looming problems, I fail to share the joy of those who anticipate with glee the demise of public education. If public education fails, we won't go back to one room schools which serve an entire community. The communities of the one room school were smaller and far less diverse than communities today. Still the one room school did serve most of the community and helped bring people together. If public schools collapse, the new computer based system will allow most children to grow up in isolation from those who come from even marginally different backgrounds. This will accelerate our slide into enclaves and special interest groups. No nation can prosper with this type of division. In the long term, even survival will be threatened.

Survival of schools, public or private, that serve the whole community is important. To survive and prosper these schools must embrace and support change, not stand against it. The American auto manufactures lost a substantial share of their old markets. They heeded the wake-up call and are learning to compete and return to prosperity. Public education is going to lose a substantial share of its clientele. If public schools heed the wake-up call, accept the changes, and rise to the challenge, they can be an important force in shaping the new education. They can also remain an important part of education. If they cling to the past, they will go the way of the horse and buggy. Alternative education is here to stay. Smaller, not bigger, is the watchword of the day.

SHIELD

In the early days of World War II, during an air attack on a PT boat in the Philippines, one gunner crouched behind the shield on his machine gun without firing a shot. When asked why he hadn't fired his weapon, the gunner responded, he was afraid he would get shot if he didn't stay behind the shield. The other crew members didn't have the heart to tell the gunner that his shield was plywood and served only to keep the ocean spray out of his eyes. The false belief in the protection from the shield had actually increased the danger to the gunner by keeping him from firing back, which could have reduced his chances of being hit.

Let's imagine that the courts and legislatures treated the right to bear arms as they treat the right to a free press. And that they treated the right to a free press as they treat the right to bear arms.

There are many false shields in our world. It may come as a great shock to many to hear that the Constitution of the United States is one of those false shields. Those who believe that the Constitution protects their liberties are as deceived as the gunner on the PT boat. Eternal vigilance always has been, and always will be, the price of freedom.

The freedom that exists in our country continues because people support and demand it, not because it is enshrined in a 200 year old document. The only strength of that document is the people who stand behind it. The Constitution is important as a symbol. As law it means little or nothing. The most it does is slightly slow changes in the law. Without the constant support of freedom by the people, the Constitution isn't up to the job. With the constant support of freedom by the people, the Constitution makes no difference.

The Supreme Court justices "interpret" the Constitution however they please, for so long as they believe voters will let them get away with it. It would take a book to fully explore the effects of the Constitution and the limits on its power. For now I must settle for considering one example.

The Constitution "guarantees" the right to a free press and the right to bear arms. Let's imagine that the courts and legislatures treated the right to bear arms as they treat the right to a free press. And that they treated the right to a free press as they treat the right to bear arms. We would be living in a different world.

We would have almost no prior restraints on the carrying of guns. Anyone of any age could carry any gun, any place. If a person with a gun injured or killed someone, attempting to imprison that person would raise the question of whether it would have a chilling effect on the bearing of arms. If imprisoning the gunman would discourage others from bearing arms, the courts would not allow the imprisonment. In other words, murder by firearms would be governed by the laws that now apply to liable and slander.

The other side of the coin would be that the "free" press would struggle under the restrictions now imposed on gun ownership. Publishing would require a government permit and a five-day waiting period for a background check. Permits for publication of small, concealable papers would rarely be granted and mainly limited to government agents. People could be restricted to buying one such newspaper each month. We would ban the carrying of those concealed newspapers. Large papers would be less restricted, except that any type of publication deemed exceptionally dangerous would be banned.

I am not claiming that guns and the press have to be treated equally. I am only pointing out that the Constitution did treat them equally. If it were the Constitution that protected these rights, they would still be equal.

If you are concerned about your freedoms, come out from behind the Constitution and stand up for those freedoms. The Constitution isn't protecting the freedoms of Americans. It serves only to lull us into a false sense of security. A dead document, no matter how revered, can't preserve individual freedom. Living people can.

THE HISTORY OF MILK

Back in the good old days (1940s) buying a gallon of milk meant bringing home four, one quart bottles, unless the buyer had the fortitude to carry 16 half pint containers. Why hadn't milk found its way into gallon jugs?

Let's try multiple choice for the answer to this perplexing question. A) In those days consumers weren't up to handling gallon jugs. B) The technology for putting milk in gallon jugs was yet to be developed. C) It was illegal to sell milk in gallon jugs. D) None of the above. E) All of the above. For those who are not history buffs, the answer is C. This was not the only milk law of the day. Unless the white stuff contained at least 3.5 percent butter fat, it was illegal to call it milk.

What does it all mean? For one thing, the history of the Milk family shows that dumb laws are not a recent invention. Another point worth noting is, dumb laws do sometimes get repealed. Considering the number of dumb laws on the books today, the mere demonstration that such laws are not immortal is encouraging.

There was one size of container (Except for those occasional half pints.), and there was one kind of milk. All milk came in quart bottles with cream floating on top of it. This made buying milk very simple.

Then came homogenization. For a penny or two more a quart you could buy milk with the cream crushed so it didn't float on top. For some reason homogenized milk usually came with vitamin D added. I'm not sure why. Perhaps crushing the cream killed or injured the vitamins so they had to be replaced. Whatever the reason, decision making had arrived in the world of milk purchases.

A short time later the lawmakers relented to pressures from certain milk bottlers and repealed the laws that protected consumers from gallon jugs. It wasn't long until the 3.5 percent law perished too. It almost perished anyway. In Michigan a product known only by the surname "Milk" must still contain 3.5 percent butter fat. Other members of the Milk family must proclaim their given names, such as "2 percent."

Now there are enough members of the Milk family to make the cooler at the supermarket look like a family reunion. Simply going to the store and asking for milk would be like calling Information (Excuse me, I mean "Directory Assistance.") and asking for the Smith's phone number.

There were other laws in milk land. The sale of colored margarine was prohibited to protect the market for that milk byproduct commonly called butter. The legal definition of milk still in effect in Michigan is a curiously worded statute. Milk is a substance removed from the cow's udder at least twice a day. If the cow isn't milked twice a day, the substance you obtain isn't milk – Not in Michigan anyway.

Since discovering this statute I have the urge to milk a cow just once a day and see what I get. Might it be orange juice or cola? If I transported it to another state where this law didn't apply, Would it magically turn into milk? This could provide a means for knowing when I crossed the state line.

There is another interesting possibility. If the substance obtained by milking only once a day isn't milk, Do Michigan's laws governing the production and sale of milk apply to that substance? Could a wily producer evade all those laws simply by cutting back on the milking schedule? Could this product be sold as, "It's Almost Milk."

What does it all mean? For one thing, the history of the Milk family shows that dumb laws are not a recent invention. Another point worth noting is, dumb laws do sometimes get repealed. Considering the number of dumb laws on the books today, the mere demonstration that such laws are not immortal is encouraging. On the other hand, all these dumb laws that have gone to meet their makers were state laws. Do dumb laws spawned in Washington face the same risks?

Then again, What if those milk laws weren't so dumb after all? It occurred to me that the fabric of our civilization started ripping apart about the same time milk showed up in gallon jugs. Could it be that all we have to do to cure the ills of society is put milk back in quart bottles as God intended?

IT WASN'T THE GRINCH WHO STOLE CHRISTMAS!

During early November I hurried through a store heading toward the oil filters. A sign that inadvertently caught my eye kicked me in the stomach. What was the traumatic message? Santa Claus proclaimed, "Merry Christmas." I already knew Christmas approached. This premature Christmas message only drove home the closeness, and the inevitability.

Is it any wonder that we as a nation wallow in seeming insoluble problems? Any society that can turn a holiday of goodwill and joy into a time of annoyance and unpleasantness is capable of screwing up just about anything. Studies show that more people commit suicide at Christmas time than any other time of the year. Perhaps the patron saint of Christmas should be Kavorkian rather than Nicholas.

> **What went wrong? Christmas started out as Christian holiday of joy and good will. Most of the activities associated with Christmas today have about as much religious significance as a dead skunk on the road.**

What went wrong? Christmas started out as a Christian holiday of joy and good will. Most of the activities associated with Christmas today have about as much religious significance as a dead skunk on the road. Christmas is the season when millions of people feel compelled to spend tens of millions of dollars they don't have, to buy gifts that for the most part the recipients don't need or even want. Christmas is the season for desperately trying to fulfill all those "obligations" before it's too late. If we are truly honest with ourselves, How many who at least pretend to enjoy Christmas don't heave a sigh of relief when it's over?

Someone on the radio explained that Christmas shopping should be a year around task, not something put off until December. He told how he liked to keep an eye open for appropriate Christmas gifts all year. Then the job was done by December when the real madness flows across the land. While I'm spending my entire year Christmas shopping, Why not go to the dentist every week too? Summer Christmas shopping fits right in. Did you ever try buying a swim suit in summer, or a winter coat in winter? Christmas goods are about the only thing left that we can buy in season.

In the not so distant past, selecting Christmas gifts was easy. Everybody needed just about everything. They would appreciate whatever you gave them. The only fly in the ointment was that most people didn't have much money to shop with. Now people have the money, so they buy what they really want for themselves. Unless you are prepared to spend a fortune, it's going to be a real challenge to buy most people something they want and don't already have. Maybe it really is the thought that counts. Don't tell that to the merchants. A penny for your thoughts doesn't fill the cash register.

Why do we let the TV huxters and slick promoters define Christmas? I am reminded of the story of two young children. One was from a Christian family and the other wasn't. The Christian tried to explain to his friend what he missed by not celebrating Christmas. The friend replied that his family did celebrate Christmas. On Christmas morning the entire family went down to his father's store, looked at the empty shelves, and sang, *What a Friend We Have in Jesus*. Did Jesus command Christmas shopping?

In November '94 the voters shouted out a message; "We are sick of what our government turned into. No matter how many slick adds you run telling us it's good – we know it isn't." How about doing the same for Christmas? Tell the merchandisers and huxters to pack it in and get lost. Who says we have to buy all the children's toys in December at premium prices? Why not shop around and buy whenever the price is the best? Why buy summer things in winter to set around and frustrate the recipient until spring? Who says we have to do all the Christmas rituals just for the sake of doing them? Unless we see a benefit, unless we really enjoy it, Why do it?

Why can't everyone define their own Christmas celebration, and let others do the same? For so long as we let Christmas be the season for keeping up with the Joneses and the Home Shopping Network, we will continue to do more Christmas and enjoy it less. Would it be so wrong for Christmas to be a season of relaxation and happiness, rather than a time for nervous breakdowns and suicides?

CHRISTMAS SEASON

When I was a bit younger, Christmas was truly a season, not just an event preceded by weeks of torture inspired by the commercial interests. The commercial interests were there. Perhaps it is my imagination, or just bad memory, that they were then less intense.

The Christmas season began right after Thanksgiving. A very tangible event marked this beginning. The teacher passed out the parts for the Christmas program. Every student from kindergarten through eighth grade was assigned a recitation or part in a play. We devoted the next four weeks to learning and rehearsing these parts, along with Christmas songs. I enjoyed being in the program. It did nothing to dampen my enthusiasm, that as we moved closer to the big day we spent more and more time rehearsing, and less and less time in classes.

> **We laid the crates out across the front of the schoolroom and put the planks on top of them. The older students dug the curtains out of the cupboard and hung them on the wire at the front of the stage. Alakazam, our schoolroom became the community theater.**

The truth is, I found little merit in education when I was in school. I also hated to write. I still hate to write with a pen or pencil. These thoughts go straight into my computer. I went to school and did my assignments because I had no choice, not because I wanted to. Those among us who think we must lure children into education with fun and games might do well to ponder this thought. I continued in school for over 20 years and eventually obtained a doctorate. If education hadn't been rammed down my reluctant throat in the early years, I doubt that it ever would have happened. I disliked school – no, I hated it – right up to the end. I digress. Back to Christmas.

The next big event was the arrival of the stage. (No, I don't mean the stagecoach. I'm not quite that old.) One of the families of the neighborhood had some old planks and potato crates. Someone volunteered a truck and brought these forest products to the school. We laid the crates out across the front of the schoolroom and put the planks on top of them. The older students dug the curtains out of the cupboard and hung them on the wire at the front of the stage. Alakazam, our schoolroom became the community theater.

Never mind that if you tried doing something like that today, the teacher and school board would end up in jail for violating more laws and regulations than I could begin to describe in this space. Besides, the children would probably all be made wards of the court for their own good. Where was government to protect me when I needed it?

Among other forbidden activities, some evil person much in need of counseling slipped in a verse or two of *Hark, The Herald Angels Sing* or *Silent Night* between *Jingle Bells* and *Rudolph the Red Nosed Reindeer.* There might even be an appearance by three wise men, either before or after Scrooge and Marles. Yes, I grew up in a dark and evil age.

Somewhere along the road to Christmas we drew names for the gift exchange. There was usually a fifty-cent limit, or something like that. As the big event approached, the parents and older students completed making the costumes. There was still time for more evil from which government didn't protect us. Just before the big night one of the parents went out in his woods and cut down a big evergreen tree. After we set the tree beside the stage, the entire school rushed to decorate the guest from the forest.

Alas, the big night arrived. We pushed back all the desks and set up rows of folding chairs and wooden benches. Every parent, and some grandparents, would be there. With 45 or so students and their families, the old schoolhouse was bursting at the seams. Still, after the curtain closed on the final song, we always made room for one late arrival. The jolly old man in the red and white suit never made it for the program. I guess he was a busy fellow. His early arrival would, no doubt, have been a major distraction anyway.

St. Nick passed out the gifts under the tree and then opened his bag. He always had a box of candy and peanuts for the young performers. The best gift of all was, we didn't have to return to school until after New Year's Day. The worst part was that the month of Christmas was over. Sure, Christmas day was yet to come, but the schoolroom would be a dull and dreary place for the next eleven months. The worst part was the first day back at school in January. The rite of evicting the guest from the forest and sacrificing the community theater back into the reality of a schoolroom, wasn't nearly so much fun as the month long process of creating that fantasy world.

The Rites of Spring

Thousands of hens lay eggs in endless rows of cages. Eggs arrive at the supermarket by the truck load. It wasn't always this way. When I was a mere sprout of a boy, the arrival of the baby chicks was one of the rites of spring.

The brooder house, a solitary building no more than twelve feet square, set well behind my grandparents' garage. It was windowless, except the south side where panels of glass cloth, extending from the ceiling nearly to the floor, made the entire side a window. In this unpretentious building, little balls of yellow fluff grew into creamy white pullets. No roosters were allowed.

Preparation of the brooder house for the new arrivals began when Grandpa shoveled out the residue from last year. Grandma poured some strong smelling liquid into a big pail of water. With this brew, she scrubbed every inch of the walls and floor until no hint remained that chickens had ever crossed the threshold.

Grandma sold eggs to the store for thirty to forty five cents a dozen. She was quite disturbed the few time the egg price dropped below thirty cents. Dressed chickens sold for around fifty cents a pound. If the price of eggs and chickens had kept pace with everything else we would now be paying over $2.50 a dozen and $3.00 a pound.

It was time to spread the large bag of peat moss on the floor. For the duration of the World War II peat moss shortage, ground corn cobs served as a substitute. A layer of newspapers covered the peat for the first few weeks. The peat and papers ended at the disc of sheet metal, an old Standard Oil sign, beneath the stove. Fire safety was a real concern. The brooder house was for raising chicks, not roasting them.

The stove, an unimpressive piece of rust colored metal no bigger than a five gallon can, set in the middle of the brooder house. The smoke pipe extended up through the roof. A galvanized canopy to shelter the chicks surrounded the stove, extending halfway to the walls.

Then came the trip to town for feed and coal. A gunny sack full of finely chipped anthracite would warm the brooder house for the entire spring.

The day before the chicks arrived, Grandma lit a fire in the brooder stove. A mechanical thermostat opened and closed the damper to maintain a constant temperature. Electricity had not yet reached my grandparents' farm. The slow burning anthracite coal held the temperature near 90° until morning, even on the coldest March night.

The big day arrived. I went along when Mom picked up Grandma and Grandpa to drive them to the railroad depot across the tracks from the elevator. The train arrived with the morning mail that included a drab, flat brown box with little round holes in it. Each quadrant of the box held twenty-five peeping chicks. While the man from the post office loaded the rest of the mail into a trailer, Grandpa and Grandma rushed the chicks into the warmth of Mom's '34 Plymouth.

Back at the brooder house, Grandma released the chicks onto the newspapers. The little balls of fluff scurried to the shelter of the canopy. Soon they reappeared to peck at the food and water setting nearby.

I was always eager to visit the brooder house. When Grandma opened the door, the peeping gave way to the scratching sound of little feet rushing across the newspapers to the security of the canopy. A wide board stood on edge inside the only door, excluding the chicks from one small corner. I stood behind the board and watched as Grandma filled the feeders and water cans, and fueled the stove. I was in another world standing in that warm corner on a cold spring day.

Grandma sold eggs for thirty to forty-five cents a dozen. She was quite disturbed the few times the price dropped below thirty cents. Dressed chickens sold for about fifty cents a pound. If the price of eggs and chickens had kept pace with everything else, we would be paying over $2.50 a dozen and $3.00 a pound. Maybe Grandma didn't raise chickens in the most efficient manner. Still we lost something when the arrival of the baby chicks ceased to be one of the rites of spring.

IN PRAISE OF NOTHING

Fire is a frightening thing. When Wilber started his lawn mower, flames belched into the papers piled in the corner of his garage. Wilber wasn't a fire fighter. He ran for help. A crowd gathered.

Anyone who believes dousing the health care system with a big bucket of federal bureaucracy will cool those flames of excess cost must have spent the last fifty years with Rip VanWinkle.

A tall man with a look of determination strode from the crowd. "I'll take care of it," he announced in a sincere voice that made believers of almost everyone. The good Samaritan picked up a pail that set at the side of the garage and swung it toward the fire. "Stop," yelled Wilber. No one heard him over the applause of the crowd.

When the contents of the pail hit the fire, flames flashed through the garage. In a matter of minutes there was no garage. Wilber vented his anger at the would be fire fighter. Why would anyone throw gasoline on a fire? The stranger tried to explain. He thought it was water.

"He meant well," intoned a voice from the crowd.

"At least he did something," chimed another.

Yes, the stranger did do something. Is doing something always better than nothing? Wilber doesn't think so.

There are problems in the health care system. Costs are too high and rising at an alarming rate. Some people don't have money to pay for the services. That's about the extent of the complaints. The soaring costs have a lot to do with why some can't afford the services. The two problems merge into one. Should we burn down the entire system because of a couple of problems?

Waste and inefficiency burn within the health care system. How can we quench the flames? Controlling waste and improving efficiency are two areas where government is at its absolute worst. Remember who brought us the $500 screwdriver and the $700 toilet seat? Anyone who believes that dousing the health care system with a big bucket of federal bureaucracy will cool the flames of excess cost, must have spent the last 50 years with Rip VanWinkle.

Maybe President Clinton means well. Maybe he doesn't. Intentions don't count. No credit is due those who do absolutely the wrong thing. We should give politicians more credit for what they don't do. It isn't important that critics of President Clinton's health care plan didn't have brilliant solutions of their own. By stopping the President from burning down the garage, we bought some time to put out the fire. If the federal bureaucracy gets hold of health care, it will give a big boost to the nostalgic among us. Where we are right now will soon be the good old days.

There are ways government can prod the health care system toward more efficiency. Just like with a new drug, these prods should be tested before being prescribed for the entire nation.

There are ways government can prod the health care system toward more efficiency. Just like with a new drug, these prods should be tested before being prescribed for the entire nation. Also, just like with a sick body, the health care system must heal itself. Government may be able to provide a few useful drugs to aid that healing. Government can't take over the healing any more than a physician can take over the healing of the patient's body. For those who still see more bureaucracy as the answer, ask Wilber about his garage.

P.S. The above was first published during the dark days of early 1994, when even most optimists saw some form of socialized medicine as inevitable. We did stop the Presidents Clinton from burning down the garage. The roof still leaks. If we don't fix those leaks, the health care garage will rot. Rot is slower than fire. In the end the results are the same. Ignoring the problem won't make it go away.

A Parable for Our Times

The village lay nestled in the mountains of a distant and mythical land, accessible only to the most determined adventurer. Alas, television arrived. The curious villagers clustered around the small black and white screen in the marketplace. Soon the village elders had sets of their own. It wasn't long until almost every family owned a TV set.

Then came the problem. Pictures disintegrated into wavy lines or disappeared altogether. The distressed villagers summoned the repairman to fix the sets. He expected to be paid. Sometimes he didn't get paid. This payment thing inconvenienced both the owners and the repairmen. Villagers who couldn't afford the repairmen's fees were sometimes deprived of the pleasures of television. Unpaid bills kept the repairmen from enjoying the prosperity due members of such a noble calling.

Then came stereo. Big screens too. The TV fee went up again to cover VCRs. This only concerned the employers who were now stuck with paying the fees.

The repairmen had an idea. For a mere pittance paid each month they could fix everyone's televisions, whenever they broke. The repairmen hired a bright young man named Hue to put this plan to work. It took a bit of selling but Hue was up to the job. Now all the televisions worked beautifully. The repairmen sent their bills to Hue who always had the money to pay them.

Would it be that the tale could end now with everyone living happily ever after. Alas, there was trouble in Camelot. TVs wore out. Even the best of repairmen could not fix them. Hue came to the rescue. For a slightly greater pittance each month, he would pay for replacing the worn out sets.

The villagers might have enjoyed eternal bliss, except for the invention of color television. The villagers demanded that the repairmen make the old sets show color. When the villagers found out that wouldn't work, they demanded new color set. Hue explained, he only paid to replace broken sets. The villagers considered that most unfair. Hue didn't want to annoy the villagers. He agreed to pay for new color sets. Of course the monthly fee went up. It was now far more than a pittance. Many villagers went along with the higher fees. It was better than having to buy a television set. Others decided the plan now cost too much.

Villagers deprived of color TV grew ill tempered. This affected their work. The employers didn't like that, so they agreed to pay the television fees. Soon every villager had a color set. One set wasn't enough. Again the fees went up. All the happy workers got a second set. Then came stereo. Big screens too. The TV fee went up again to cover VCRs. This only concerned the employers who were now stuck with paying the fees.

The prices of television sets kept going up. Few people even asked the cost. Just send the bill to Hue. There were big screen televisions in every room of every house – sometimes two of them. A spare is always a good idea.

The employers who now paid the television fees complained. No one listened until the employers went bankrupt. Then it was too late. Production ground to a halt. The villagers starved.

Someday mythical archaeologists will come to excavate the site of our mythical village. They will find remains of houses filled with skeletons lying in front of big screen television sets. Never will they find two skeletons in front of the same set. Perhaps the most curious find will be the monument on the hill overlooking the village. Will the archaeologists understand this tribute to Hue Cross, the man who brought unlimited free television to the villagers?

WHO SHOULD OWN YOUR HOUSE?

For many, home ownership is a measure of success. It's a big part of the American Dream. Some people choose not to buy a home although they could. Others rent because they have yet to achieve an income adequate to buy a house. Still, even renters are responsible for finding and choosing a home.

There's no such thing as a typical home. Houses, apartments, trailers, condominiums – as varied as this list is, the differences between types of homes are less than the variations within the types. A penthouse apartment has far more in common with a single family mansion than with a one bedroom flat.

Why on earth would even the most foolish among us even consider replacing our system of personal housing with employer furnished housing and all the problems it would create? Why would employers ever start down this road? Why would employees follow? Why would government favor such foolishness with tax incentives?

Economic considerations are a big factor in selecting a home. Whatever the financial circumstances, the family home is a bright reflection of the needs and desires of the family. As circumstances change the home changes, whether through remodeling, building an addition, or moving to a new abode.

A few people have little choice in the selection of their homes. The home comes with their job. This was more common in an earlier era. Perhaps the largest remaining vestige of employer furnished housing is found in the churches that often still provide a house for the minister or priest.

Not having to make monthly payments, not having to fix the roof, not having to sell and buy a house when moving, has its advantages. These benefits come with a price. The employee gets the house he gets, whether or not it's suited to his needs or taste. Sure, he doesn't have to accept the employment. It's difficult enough to find a job without having to consider the house that comes with it. Tying housing to employment wouldn't fly well with most Americans. Especially, if each employer provided only, one size fits all, row housing. Housing designed by the employer with his costs and needs in mind is unlikely to meet the individual needs and tastes of most employees.

There are other problems. When the job ends the employee has to move. If he has another job and another house, that's not much of a problem. What if there is no new job? The worst time to be pushed into seeking a place to live is while unemployed. To make matters worse, if employers owned most of the houses, the housing choices available to the unemployed would be scant.

So why waste time worrying about a problem that doesn't exist, and is unlikely to exist anytime in the foreseeable future? Maybe we could get employees to accept company housing. Suppose housing was provided as a fringe benefit? The employee can refuse the house. Refusing the house does not increase the employee's pay. Suppose government gets into the act. What if we amend the tax code so that employees don't pay any income tax on company furnished housing? The employee would still pay income tax on any money he earned to pay for his own housing.

This is the road we have followed with health care. Yes, it was insane. Why we did it is not of great importance. What we are going to do about it now is of utmost importance. There is a health care problem.

Chances are that within a generation most people would be living in employer owned housing. That doesn't mean everyone would be happy. Loss of a job would be far more traumatic than now. Even when job loss wasn't a problem, one or two sizes fits all housing would leave much to be desired. Many employees would take poorer care of the houses than they would if their investment or security deposit were at stake. Employers would seek ways to prevent abuse of houses and keep costs down. Employers would soon be more than ready to wash their hands of company owned housing.

Once hooked on company housing, employees would seek ways to get the companies to furnish more desirable housing. One avenue for seeking improved housing would be appeals to government. If government took over the housing, that would at least solve the problem of housing being tied to the job.

Why would even the most foolish ever consider replacing our system of personally provided housing with employer or government furnished housing, and all the problems it would create? Why would employers start down this road? Why would employees follow? Why would government favor such foolishness with tax incentives? I would like to believe that in a sane world this would never happen.

This is the road we followed with health care. Yes, it was insane. Why we did it is not of great importance. What we are going to do about it now is of utmost importance. There is a health care problem. We now hear little about it from Washington. After 1994's brutal battle to fight off that final step to replace company owned health care with government owned health care, the politicians are too beaten down and exhausted to bring up the issue. Also, some problems that will reappear in the next recession aren't so widespread now.

The only solution is to go back to privately furnished and "owned" health care. It wouldn't be difficult to do. It is ironic that the federal government, which tried to nationalize everyone's health care, now provides it employees with one of the better systems in the nation. Even government can do something right once in awhile. The federal government recognizes an array of insurance plans from which the employees can choose. The government makes a contribution toward cost. The employee pays the rest. The more expensive the plan the employee chooses, the more he pays. I'm not suggesting that this plan is ideal. It is superior to what most private employers offer.

We must amend the tax code so that money employees spend on health care is not any more subject to income tax and other taxes than are employer contributions. Until consumer "owned" insurance stands on the same tax footing as employer provided insurance, the company owned insurance will have an unfair advantage in the marketplace.

What are some of the benefits from having individuals own and pay for their insurance? Changing jobs would not be a problem because insurance would not be any more tied to your job than is your home or automobile. Individuals' needs and desires regarding health insurance would vary as much as their needs and desires regarding housing. This may not be obvious in our one size fits all world. It would soon become apparent in the world of choice.

With the consumer now a player in the health care cost game, competition and accountability would follow. Costs would fall. Just like with auto insurance, consumers would soon discover that they could save a bundle by having higher deductibles and paying the small bills themselves. There are at least two arrows in this quiver. The overhead cost at both ends of the billing would be greatly reduced. The consumer would do a far better job than an insurance company in making sure he is not over billed. Insurance companies cannot afford to police small bills. It's cheaper, at least in the short run, to pay over billings. The more the insurance company will pay, the higher the bills get.

The only solution is to go back to privately furnished and "owned" health care. It wouldn't be that difficult to do.

Medical insurance markets would no longer be dominated by a few large purchasers, with little interest in the quality of the product. The market would be open to an all but unlimited variety of insurance plans with deductibles and coverages to fit varying needs. The insurance companies that provided the best services at the best prices would be rewarded. Consumers would reap the benefits of the efficiency and cost cutting. The consumers would soon discover that their efforts toward keeping down costs would benefit them. Consumers could form cooperatives and other buying groups to negotiate favorable insurance coverage and prices.

What about those who buy no insurance and then go whining to government for a bailout? If necessary, a small payroll deduction could provide minimum coverage for all employees. If the employee refused to select coverage and use this money for insurance, the employer or an appropriate government agency could assign the coverage to an insurance company. Assignments would be made at random to one of the low bidders. Any room for bureaucratic discretion would breed corruption and monopolies. Sure, this won't take care of those who are unemployed. Neither does the present system. Lower medical costs will benefit the uninsured, thus improving their condition. The mere fact that consumer "owned" health insurance won't solve all the world's problems, is no excuse for failing to implement it for the benefits it will provide.

If we don't go back to consumer owned and controlled health care, it's only a matter of time until the feds nationalize the present mess into socialized medicine. That will be to health care what government ownership of all the houses would be to housing. If you would like to take a peek at what national health care will end up looking like, visit some big city housing projects. Don't forget your bodyguards.

WANT

Americans want a lot. Who doesn't want more income, more leisure time, more safety, more security, a bigger house, a better car, a better job, better education, bigger and better everything? The people who are satisfied could hold their annual convention in a phone both, with enough room left for Superman to do his quick change. Want and dissatisfaction aren't bad things. Want and dissatisfaction are the great motivators. Without them we wouldn't have achievement or progress.

Dissatisfaction is a two edged sword. When supported by effort and tempered with realism, it can propel us to great heights. When the effort and the realism are missing, dissatisfaction yields only frustration. Frustration yields failure, violence and destruction.

For continually increasing millions of Americans, want generates frustration rather than effort. If there is any one explanation of what's happening in America today, this is it. In life you must play the hand you are dealt. Screaming "misdeal" and demanding a new hand doesn't work. Moaning "it isn't fair," doesn't help either. If someone told you life would be fair, they deceived you. It never has been and never will be. Get used to it. Those who take the ball and go home because the game isn't fair are guaranteed losers. For each individual the struggle out of life's swamps begins wherever they are at.

Millions of Americans wallow in failure because they fail to seek and seize the little opportunities that are available. If they can't go straight to the major leagues they won't play.

A newspaper article reported about an individual who lost his job. He turned down another job because it wasn't good enough. After going through a government retraining program he ended up taking essentially the same job offered in the first place. It appears that he learned something. Take the opportunity that's available. This is sometimes known as "the bird in the hand is worth two in the bush" principle.

Millions of Americans wallow in failure because they don't seek and seize the little opportunities which are available. If they can't go straight to the major leagues, they won't play. Unemployment benefits, welfare and a great morass of other government programs make it easier for millions to overlook the little opportunities. Unless I've completely misread history, during the great depression the unemployed didn't turn down opportunities just because the opportunities weren't big enough. This "will do" attitude carried us through World War II and propelled us into the prosperity of the fifties. We still live on the residual of the strength generated by the harsh times.

Does anyone have the audacity to say, there are fewer opportunities today than during the 1930s? Attitude, not lack of opportunity, is the problem. Many people are unable to avail themselves of the opportunities because they have not prepared themselves. This lack of preparation springs from bad attitudes.

Many make the excuse that today's world demands so much more of people, that a person must have much more knowledge and ability to survive. This myth is based on ignorance. We live in the age of the specialist. Most jobs involve doing a few little things over and over. Those who farmed or ran family businesses a century ago had far broader responsibilities. Earning their livelihood required far more skills and knowledge than most people use to earn a living today. A woman managing a household and caring for a family in the nineteenth century needed far more skill and ability than most career women need today. The mere fact she didn't learn those skill in college didn't make them any easier or less important.

The one thing that hasn't changed is the need for a "can do, will do" attitude. Job success is far more a matter of attitude than of ability or training. Employers want workers who are dependable and willing to work. Those with good attitudes can learn the skills needed to work their way up to better paying jobs. Those who wait to start at the top, and demand to do it on their own terms, are on the road to nowhere.

I long ago lost track of the number of people who tell of quitting jobs because they didn't like the boss or the working conditions. Most of the quitters didn't have another job when they quit. Wherever you go, you will find some abrasive, unreasonable people. Conditions will seldom, if ever, be ideal. Those who succeed learn to make the most of the conditions in the real world, rather than using those conditions as excuses for failure.

No program from Washington is going to cure the ills of America. Only an attitude adjustment on the part of millions of Americans will do the job. Government is adjusting the attitudes in the wrong direction. This is why government can never solve the problems. The answer is for everyone to seek out and seize whatever opportunity there is. No opportunity is too small if it's a step in the right direction. For those at the bottom there is only one direction. Individuals must pursue even the smallest opportunities like their lives depended on it. In the long run everyone's life does depend on it. Those who wait for opportunities to come to them are in for a long wait. For those who demand ideal opportunities, the wait will be even longer.

Prostitutes, Margarine and Handguns

What do prostitutes, margarine and handguns have in common? Perhaps it's just as well not to consider all the possibilities. What is the common thread running through these unlikely companions?

In the beginning there were prostitutes. If not in the very beginning, they at least arrived on the second bus. Somebody decided prostitution wasn't a good idea. Not everyone agreed. So began the endless struggle to rid the world of prostitutes. Those opposed to prostitution strove mightily to cut off the supply of prostitutes. Social pressures, laws, appeals to Deity, all failed to rid the world of the ladies of the evening. The supply still equals or exceeds the demand. Close down one street corner – they pop up somewhere else.

Sometime between the arrival of the first hooker and the discovery of colored margarine, the Chinese invented gunpowder. Gunpowder led to guns. Did they call it gunpowder before guns were invented? Were guns invented because it was rather silly to have gunpowder without guns?

In recent times another scourge threatened to spread across the land. The impassioned pleas of those alert to the peril fell on sympathetic ears of their legislators. The evil threatening to destroy life as we knew it was banished to the nether land to dwell with prostitutes, pimps and bootleggers. Banning the sale of colored margarine made the world safe for butter and the contented cows that produced it. The consumers' dollars would continue to flow to dairy farmers as God intended.

Did the scheme work? Depends on how you look at it. The purity of the store shelves was inviolate. Only the whitest of margarine dared show itself. Never underestimate Yankee ingenuity. Clever scientists discovered that white margarine plus yellow food coloring equaled colored margarine. They even developed the technology by which the most unsophisticated consumer could do the mixing. If the goal of the anti margarine campaign was simply to prevent the sale of the colored stuff, it succeeded. If the goal was to keep colored margarine off the tables of America, the scheme failed miserably.

Sometime between the arrival of the first hooker and the discovery of colored margarine, the Chinese invented gunpowder. Gunpowder led to guns. Did they call it gunpowder before the invention of guns? Were guns invented because it was silly to have gunpowder without guns? Like with prostitutes and colored margarine, not everyone found guns to be an unmixed blessing.

Those who believed that the wrong people were being perforated with bullets, attempted to make sure that only those who would aim in the proper direction would have guns. The fly in the ointment is that the people we least want to have guns are the ones bothered the least by the simple fact guns are illegal. There is only one way to keep these people from getting the guns they want. We have to make it impossible for those bad guys to get guns. We would have to eliminate all sources of supply.

A passion for cutting off supply burns deep in the human soul. When we don't like something, we want to get rid of it. It hasn't worked with prostitutes. It didn't work with colored margarine. Prohibition wasn't a rip roaring success either. In the history of the human race how many evils were squelched by cutting off the supply? If I can think of one and you can think of another, that will make two.

Most opponents of prostitution begrudgingly admit that prostitution will not end. The battle to stop prostitution continues more because the warriors would rather fight a losing battle than admit to defeat by evil. There is a strong case that the harm from prostitution could be reduced by legalizing prostitution. Still for moral reasons, many, if not most, people are not persuaded that we should tread that road. They may be making the right choice.

The moral issues involved in prostitution do not apply to guns. We don't have to fight a no win war against handguns just to preserve our principles. So, why are so many otherwise rational people obsessed with the futile effort to cut off the supply of handguns? The only thing that will end that supply is cutting off the demand for handguns. If criminals don't want handguns, no one will bother supplying them.

The only way to deal with the handgun problem is by directing our efforts at the users. It will be a long, slow process. Success is possible. Attempting to cut off supply only diverts energy from doing something that could help. Whether you like it or not criminals will have guns for as long as they want them. The answer to gun crime lies in cutting off the desire to commit crime with guns, not in cutting off the supply of guns.

Some may say that ending the criminals' desires for guns will be difficult. Some may even believe it is impossible. It will be difficult. It is possible to reduce the number of people who will seek to use guns for criminal purposes.

In many nations the desire to use guns for criminal purposes has not risen to the level it has in this country. This alone shows that a lower level of desire to use guns in crime is possible. If the desire for guns in these other countries was as great as it is in our country, the guns would be there to fill the demand.

It's time we laid to rest the myths and delusions about the wonders we can accomplish with gun control. It's time we direct all our energies at the misusers of the guns. Attitude adjustments take time. It's the only hope we have for success.

CAN WE TAKE A BITE OUT OF CRIME?

Mark Twain quipped, "Everybody talks about the weather but nobody does anything about it." Based on results the same could be said of crime. Voters demand action on crime. Politicians make much noise and pass laws. It should be painfully obvious to all that those laws aren't stopping crime. Federal crime bills are election year gimmicks. They don't help. They often hamper the battle against crime.

It is very difficult to solve a problem without understanding it. That's why cures for cancer are so elusive. Until recently, no one had even a clue about what caused cancer. Attempts to cure cancer were like throwing darts at balloons in a dark room. When the scientists heard a balloon pop they assumed they must have done something right. They had no idea why they had succeeded. Now that scientists are unraveling the secrets of cancer, there is some promise that real cancer prevention and cancer cures lay just ahead.

Crime grows because as a society we tolerate it and let it grow. Harshness against criminals may do little to reform them. Harshness against criminals will help reform the social attitudes that result in individuals growing up to be criminals.

Most of those who lead the charge against crime do not understand their adversary any better than the cancer researchers understood theirs. The difference is, cancer researchers knew they didn't understand their adversary. They sought understanding.

It is well past time for the warriors against crime to seek to understand their adversary. For several decades we have pursued dangerous myths that increase crime rather than decreasing it.

Crime is a symptom, not a cause. Criminals cause crime. Crime does not cause criminals. We learned centuries ago that swatting mosquitos was not effective when there were more than a few mosquitos. To control mosquitoes, we had to drain the swamps. Swamp draining has fallen into disrespect. If our ancestors had preserved swamps the way some people seek to now, most of Michigan, and the rest of the Eastern Midwest, would still be uninhabitable by humans.

If we don't want our country to again be uninhabitable by humans, we best drain the swamps that breed crime. Deluding ourselves into believing that swatting criminals is the answer leads only to failure. I'm not against swatting criminals. I still swat mosquitoes now and then. Neither act by itself will ever cut off the supply.

Crime grows because as a society we tolerate it and let it grow. Harshness against criminals may do little to reform them. Harshness against criminals will help reform the social attitudes that allow children to grow up to be criminals. The emphasis on reforming criminals created an atmosphere where criminals get far too much sympathy and understanding.

I recall the story about a farmer buying a mule. The seller assured the farmer that the mule would do whatever the farmer told the mule to do. The mule refused to do anything the farmer told him. The farmer called the seller who came over to take care of the problem. The former owner hit the mule over the head with a club. Then the mule did whatever it was told. The farmer complained that the seller promised that all the farmer need do was speak to the mule. The seller replied that first he had to get the mule's attention.

We have to get the attention of criminals, and would be criminals, if we are to prevent crime. Stressing rehabilitation while winking at juveniles and other first time criminal offenders doesn't get anybody's attention. By the time an individual embarks on a life of crime it's too late to prevent crime. We are then faced with the task of catching criminals. This is the equivalent of only swatting mosquitoes after they bite someone. What we do is worse than that. For the first couple of bites we don't even swat. We just brush them away a bit.

Swatting criminals after the third big bite isn't the answer. We must start swatting after the first little bite. Hit them hard to get their attention before we attempt rehabilitation. The message that the criminal gets clobbered for his first little crime, will get the attention of not only the criminals, but also the younger children and their parents. It will take time but we can again raise a generation that will respect the lives and property of others. Holding the parents of juvenile criminals accountable will strengthen the message and improve the results.

Parental and community failure to instill proper attitudes in children is the main cause of crime.

We must root out the attitudes that breed criminals. One of the most important reasons for whacking criminals is the message it sends. If we want to send a meaningful message we must hit hard, quick and often. When the whacking doesn't begin until after years of rehabilitation and coddling fail, we send the wrong message. We don't have to hit young criminals with a cane. We do have to hit them hard with something. Parental and community failure to instill proper attitudes in children is the main cause of crime. Cracking down on young offenders, and their parents, can help shape up the parents. If the parents don't shape up, the generations of criminals will continue to roll.

NEIGHBORHOOD COPS

When wandering in darkness, seek light. For so long as I can remember government promised to lead us from the darkness of the social and political failure spreading over our land like lava from a volcano. It is easy to criticize and ridicule government's failures. Failed government programs are not difficult to find. Merely eliminating the failures will not bring success. Darkness is the absence of light. It is impossible to drive away darkness. The only answer is to bring in the light. Are there any beams of light rising from the chaotic jungle where the beast we call government dwells?

B.C. (Before Cars) police officers walked beats in the neighborhoods. They were part of the community. Only law breakers had reason to fear the officers. The automobile has separated the police from the public they are suppose to serve.

Occasionally government does something right. It may be overstating these successes to call them beams of light. They are more like camp fires glowing from clearings in the jungle. It is far better to seek these faint embers than to flail at the darkness. If we fan these sparks, some day their light may spread throughout the land.

Where is the light? My nomination for "most promising government action of the decade" is the neighborhood police programs growing in a number of cities. A police officer is assigned to a neighborhood. The officer may live in the neighborhood he serves. The officer interacts with the community rather than driving around in a car. Early results show such programs are successful.

These programs may dampen police recruitment. Many are attracted to police work by the opportunity to drive around in a fast car with flashing lights. Keeping this kind of person out of police work will be no loss.

Television news profiled another police program. Some Texas schools have their own police departments. It is most regrettable that respect for life and property has declined to where such drastic action is necessary. We must play the hand we are dealt. There is a positive side. The police become fellow human beings. Students see the police as protectors. The students who discover that the police make the schools a safer, better place, see the officers in a new light.

The success of these programs shouldn't be surprising. Neighborhood cops aren't exactly a new idea. B.C. (Before Cars), police officers walked beats in the neighborhoods. They were part of the community. Only law breakers had reason to fear the officers. The automobile separated officers from the public they are supposed to serve. Now police officers lurk behind billboards waiting to pass out tickets to anyone who happens to stray above the speed limit. Most of the public sees police officers more as enemies, than as friends. There is a wall between the community and the police. It's "us" and "them."

Police alone can't stop crime. The entire community must be involved. Until the "us" and "them" becomes "we," don't expect the crime situation to improve. McGruff can't take a bite out of crime all by himself. Everyone has to nip at the heals of crime, if we are to drive it from our neighborhoods.

A police officer in a neighborhood can cut crime. The greater benefits will be longer term. Imagine a generation of children who grow up knowing the cop on the beat as a friend and neighbor. Contrast that with the current generation that sees a cop as someone looking to give mom or dad a speeding ticket or worse.

A cop on the beat can't cover nearly so much territory as one in a squad car. Doing a small job well, is far better than failing at a big job. Police darting about in cars quickly reach the crime scene. Rarely does the officer arrive before the malefactor flees. The officer writes a report and feeds it to a computer creating mountains of statistics. Statistics don't stop crime. We need some police officers in cars. Even more, we need officers out with the people.

The most effective police force makes the fewest arrests. This doesn't mean that a police department can improve its performance simply by not arresting criminals. It does mean, the truly effective police force is the one that prevents crime – not the one that arrests tons of felons. An ounce of prevention is still worth a pound of cure.

What Should We Do With Criminals?

The criminal who changes his ways and becomes a law-abiding, productive citizen, ceases to be a problem or burden. Some among us are obsessed with rehabilitating all criminals. Others are more concerned with retribution and punishment. Punishment accomplishes nothing, unless it prevents crime. Rehabilitation usually doesn't work. What is the answer?

Like all other human characteristics, the tendencies toward irresponsibility and crime are stronger in some individuals than others. Some people are all but certain never to knowingly commit a crime. Others are equally certain to do whatever they like, without any regard for laws. Most people fall somewhere between. Those at the bottom end of the scale are the biggest problems. They are the repeat offenders. Individuals well up the scale may sometimes commit crimes – even heinous ones. Fortunately this doesn't often happen.

We should radically change the way we sentence criminals. Every convicted felon should be made a permanent ward of the state, in other words a life sentence.

Rehabilitating criminals requires moving them up the scale of responsibility to a law-abiding level. Some can be moved up far enough that it is safe to return them to complete freedom. Some can't be changed. The only way to deal with those at the bottom of the scale is to forcibly prevent the commission of further crimes. This means permanent imprisonment or execution.

We have relied on imprisonment, probation and a few executions to change the ways of criminals. In recent years we made some efforts to fill the gap between prison and minimally supervised probation or parole. Our efforts are far too meager. The amount of supervision and control required to correct the conduct of a criminal varies from infinite to minimal. When we focus on only the infinite (incarceration) and the minimal (probation and parole), most convicts fall through the cracks and back into crime.

What Should We Do With Criminals?

Individuals are in prison because they failed at self regulation. Prison and other forms of external regulation are poor substitutes for the missing self regulation. They are like casts for broken bones. Unless the bone heals itself, the cast must be permanent. Acquiring new habits takes practice. Prison doesn't provide much opportunity to practice self discipline. We must put the convict in an environment where we can nudge him along one step at a time.

Prisons must be the repositories for those too dangerous or too hopeless to ever be granted freedom. For the rest, it should be at most the first stop on the long journey to responsibility. No inmate should ever go from prison to complete freedom in one step. For many the journey can begin at a station short of prison. In prison those destined to return to the outside should be separated from those being stored for life.

Extending sentences by doing away with parole is worse than foolish. Convicts must be motivated and supervised on the journey to responsibility. Eliminating parole increases the chances the inmate will remain a criminal. There are far better ways to extend sentences. We must radically change the way we sentence criminals. Every convicted felon, regardless of age, should be made a permanent ward of the state – in other words, a life sentence. The convict could gain freedom only by earning it. He would be released from supervision only after completing the rehabilitation program, and proving his rehabilitation by a period of responsible living, say five years. Until release, any lapse by the convict would return him instantly to prison, or whatever other degree of control was appropriate. There is no way to predict how long it will be before a convict is ready for complete freedom. Some will never get that far.

Work and education release would be the first stops on the journey to responsibility and freedom. Programs such a "boot camp" could serve to prepare the inmates for the first step out. We should not treat all convicts the same. At first the convict would be escorted to work or school and watched. Those who proved worthy would get more freedom to go and return on their own. The successful graduates at this level could be placed in a dormitory type setting subject to strict rules and curfews. As they progressed the restrictions would be eased.

When appropriate, the inmates could be released on electronic tethers to begin more normal lives. Again, increased freedom could be earned by degrees. The final stop would be periodic reporting to a parole officer or counselor.

None of the programs I suggest are new or even novel. I'm sure that all exist somewhere today. What is missing is an integrated program providing all the steps up the ladder to becoming a responsible citizen. A ladder with missing steps is little better than no ladder at all. One reason given for not fixing the ladder is, we can't afford it. Is it less costly to have the convicts falling off the broken ladder and remaining in their old ways? Is this just another example of not having the money to do the job right, but having the money to do the same job over and over?

I'm not under any illusion that we can nudge all convicts into responsible living. Most convicts soon return to freedom, whether they have been nudged or not. If we can help 50, 30 or even 10 percent, we will be far ahead of where we are today. Keeping non violent offenders in rehabilitation programs outside prison will make room in prisons for those whose violent and deadly acts show them as unfit for another chance. We shouldn't wait for these dangerous felons to commit a third atrocity before we put them away.

The first step is to establish pilot programs to test and perfect the road from convict to responsible citizen. No one will know exactly what works until we try. Experience is still the best teacher – often the only teacher. We will have to make many judgments about where to start the convicts in the program. There will be many more decisions regarding when and how to increase the degree of responsibility and freedom. Some convicts will slip back a few rungs and have to start climbing again. Once we learn how to make the system work, we can expand it until it's available for all convicts who can pass the entrance examination.

We must begin giving irresponsible individuals the amount of supervision required to allow them to learn responsibility. There is one indispensable ingredient. The criminal must want to become a responsible person. Without that desire no program will work. The first step for those without the desire must be to create at least a spark of desire. The rest of the job is fanning that spark.

WHAT'S WITH GLOBAL WARMING?

I first heard about global warming when I was a wee lad back in the forties. It seems that many, many years ago the world was rather a cold place. Much of our world was covered with big sheets of ice. We now call that time the ice age. Then came global warming. The glaciers that once reached to near the Ohio River retreated to the Arctic Region.

In recent times the world's temperature has been reasonably stable. There never was any consensus about what that temperature might do in the future. I recall several articles on world climate published in the fifties. Some writers advanced the theory that the ice age had not ended. The world was still warming. The glaciers were still in retreat.

If all the worrywarts who are in a tizzy over global warming had lived while the Upper Midwest was still buried under a sheet of ice, Would they have been in a panic when the glaciers started to melt? Suppose they had been living the life of Eskimos down in Alabama just beyond the glacier's reach. Would they have feared that waters from the melting ice would wash them away?

Some scientists enthusiastically urged the opposite point of view. The new ice age had already begun. The glaciers were on the march toward us. As if that wasn't confusion enough, one scholar pointed out that the past fifty years had been unusually warm. I don't recall if he said the whole world had warmed up or just North America. He hastened to note that such warm periods were rare and brief. We should prepare for the return of the normal, colder weather of the eighteen hundreds.

The advocates for each point of view trotted out statistics to support their conclusions. I never figured out who was right. I was certain though that they couldn't all be right.

It wasn't until the seventies that the Chicken Littles and Nervous Nellies of the world banded together and found the answer. The world was warming. We were causing the warming by burning too much oil, coal and stuff. Disaster lay ahead, unless we mended our ways. The world would warm. The oceans would rise. Untold evil would be visited upon our world.

Don't mind that there is no proof the world is getting warmer. The whole global warming scare springs from computer models showing that the world will warm up. The nice thing about computer models is, they always tell what the programmer tells them to tell. A rumor may eventually come back to the person who started it. The return of the rumor does not prove its truth, even when the rumor is circulated through a computer.

We are expected to overlook the total lack of evidence that our life style is causing something that may not even be happening. In many circles it is an article of faith that the human species is causing catastrophic global warming. On the other hand, there are those among us who vigorously assert that we need not worry because the world isn't warming.

Allegedly, the cause of global warming is increased carbon dioxide in the atmosphere. There is no doubt that carbon dioxide has increased. There is only one proven effect of this small increase. It makes plants grow better. This means we can produce more food for the ever growing population. The increase in carbon dioxide is buying us some time to control world population before starvation becomes rampant. If the carbon dioxide is also warming the world, we may well be looking at a trade off. Choose between a warmer world and starvation for millions or billions of people.

I have yet to hear anyone attempt to refute the basic assumption that global warming is a bad thing. If all the worry warts who are in a tizzy over global warming had lived while the Upper Midwest was buried under a sheet of ice, Would they have panicked when the glaciers started to melt? Suppose they had been living the life of Eskimos down in Alabama just beyond the glacier's reach. Would they have feared that water from the melting ice would wash them away? Would they have fretted that the warming would kill the seals and walruses essential to their survival?

WHAT'S WITH GLOBAL WARMING?

The melting of the glaciers changed the world. The oceans rose. Some places became less habitable. Extinctions did occur. How many want to go back to the world of the ice age? Would the global warming cultists be any less apprehensive if they believed the world was cooling instead of warming? Don't bet on it.

Isn't it amazing that the world has managed to achieve its perfect temperature just at that second in history when we happened along? The people worrying themselves sick over global warming just can't stand change, whatever it is. They always see the negative side. They make it sound like there is no positive side to global warming. How can a resident of the Upper Midwest go outside on a January day and be against global warming?

I hope the global warming people are right. I hope the world is warming. I'm not holding my breath. Even if the warmer world is coming, it won't be here for a long, long time. Sure there is a down side to global warming. Name something that doesn't have a down side. Isn't it just possible that some future generation will look back at today and wonder how we survived in such a cold world?

When Extortion Prospers, None Dare Call It Extortion

"Texas community says no to new jobs." This was the lead for a news report. The report implied that the community must have lost its collective mind. The report went on to tell how the community refused to approve a tax abatement for Apple Computers. Refusal to grant the tax abatement was the equivalent of telling Apple to get lost. No subsidy, no business.

The news report focused on the reasons alleged for denying the tax abatement. Supposedly the local government denied the vital abatement because some of those making the decision didn't like Apple's fringe benefits. Apparently Apple provided insurance benefits for their employees' live in homosexual friends. The report implied that denying the tax abatement was a politically incorrect rejection of homosexuals. Remember when news was news and commentary stayed on the editorial page?

We want to cut government spending at all levels. Let's start with business subsidies – all of them. We won't have to level the playing field if we quit digging holes in it with subsidies.

The report never questioned that a business was entitled to a tax abatement for building a new facility. Has extortion and bribery became so much a part of the American way that they are now rights? When a business demands payment for locating in your community, Must you just give in to the extortion and say, "How much?"

How did this deplorable condition come to be? As I recall it got its big start after World War II. In the beginning there were bribes. Various communities, particularly those in the South, started dangling juicy tidbits to lure businesses to the neighborhood. It worked. Other communities responded with tidbits of their own to keep business.

Soon businesses demanded compensation for coming to your community or staying in your community. Simple bribery had evolved into extortion. This extortion now reaches for anointment as a God given right.

We condemn congressmen, generals and bureaucrats who accept bribes. We also condemn the contractors and other favor seekers who pay the bribes. How do these payments differ from those given to lure business? Sure, one is legal and the other isn't. This is no answer. We could either make both legal, or both illegal. This does nothing to answer the real question.

In either case the payment is to influence a decision maker to do something that doesn't make economic sense. Why must a business be paid to locate in a community? Only one reason. The business would be better off (meaning it could make more money) if it located somewhere else. The community rewards the business for making a wasteful decision.

If America is to compete in world markets, we must start making business decisions for sound business reasons. This should start with businesses choosing the most economical location rather than looking for the biggest bribe. The cost of the bribe is just as much a cost of production as money expended by the business. The business may not pay that cost. Still that cost is a part of America's cost. In the big picture, that is the cost that counts.

American businesses extol the virtue of free enterprise and competition. Let them practice what they preach. If business leaders really want free enterprise, let them get their hands out of our pockets and practice free enterprise. If business decisions are made for sound economic reasons, we will all be better off. We want to cut government spending at all levels. Let's start with business subsidies – all of them. We won't have to level the playing field when we quit digging holes in it with subsidies.

Businesses will whine and snivel. You will hear cries that the sky is falling. The executives will scream that life as we know it is about to perish from the earth. There's nothing new about this. We who live in Michigan remember how the bottle deposit law was going to destroy Michigan if it passed. Is Michigan still here?

The calamities will not come to pass. Sure, a few businesses will fail. Businesses that are so inefficient or unnecessary that they can't survive without subsidies should fail. It is this survival of the fittest that makes the free enterprise system strong. New businesses will rise on the ashes of the fallen ones. We will all be better off.

Business subsides are as addictive as any other form of welfare. They are like cocaine or feeding the birds in the winter. It is far easier to start than to stop. There will be withdrawal symptoms from ending the addiction. It would have been far better to never have started. Unfortunately that ship has already sailed. The only choices now are to end the addiction, or continue the sickness. If we continue the sickness we all shall suffer. In the end, the disease may prove to be fatal to our economy and our nation.

INTELLIGENCE, KNOWLEDGE AND WISDOM

This is the first installment of an election eve trilogy. Buckle your seat belts and hang on.

Is it redundant to say a person is, "intelligent, knowledgeable and wise?" On the other side of the coin, What about, "stupid, ignorant and foolish?" What about combining terms from both lists? Is "intelligent fool" an oxymoron? No, an "oxymoron" isn't a dimwitted bovine, and has nothing to do with impugning someone's intellect. It only describes two words that don't belong together, such as "truthful politician."

Most who acquire wisdom break out of the narrow confines of a specialty. They sample the broad spectrum of life. They see and understand the puzzle, not just one piece. The melon from which we cut our experiences today may be far larger than in the past. Unfortunately most people are cutting a far smaller slice for themselves.

Intelligence can be compared to a bottle. The bigger the bottle, the more water it will hold. The size of the bottle tells us nothing about what's inside. Likewise measuring intelligence tells nothing about how much knowledge is in the mind. A small bottle may contain more water than a big one. Which would you rather have on a trip across the desert, a big, empty bottle or a small, full one?

Having a big bottle full of water in the desert means little if the traveler knows not how to use it. If the traveler takes a bath with the water, he may still die of thirst. Likewise, intelligence and knowledge mean little if not accompanied with the ability to use them. Wisdom is the ability to use knowledge. We sometime call it good judgment or common sense. Few things are less common than common sense.

The absence of light is darkness. The absence of wisdom is foolishness. Thus, an intelligent, educated person can be a fool. It is easy to find examples of intelligent, educated fools, particularly in

Washington. It's their Mecca. They worship at their shrines sacrificing the tax dollars and freedoms of the inhabitants of our land. Until the voters learn to judge political candidates on their wisdom, rather than their intelligence and education, we will continue to elect educated fools to public office. We will also continue to pay the price.

Why is wisdom so rare? Wisdom comes largely from experience. Experience isn't measured by the years a person has lived. One year's experience 30 times does not equal 30 years of experience. We live in a world where accumulating knowledge is relatively easy and greatly encouraged. Might I suggest that wisdom is far less common today than 100 years ago.

Consider a 1000 piece jigsaw puzzle. A scholar studies one piece of the puzzle. He measures and then writes an equation that, when plotted on a graph, exactly represents the shape of the piece. He figures out the exact chemical analysis of every shade of color on the piece. He also records the spectrograph of the light reflected from every point on the piece. The study goes on. Our expert writes a book about his piece of the puzzle. What does our expert know about the whole puzzle and how to put it together?

Wisdom, like putting a puzzle together, requires broader experience and knowledge. In our specialized society the acquiring of wisdom is far less likely than it was in the simpler world of the 1800s. Most who acquire wisdom break out of the narrow confines of a specialty. They sample the broad spectrum of life. They see and understand the puzzle, not just one piece. The melon from which we cut our experiences today may be far larger than in the past. Unfortunately most people are cutting a far smaller slice for themselves.

Unless we choose wise leaders, the foolishness will continue. Unless voters are wise, How will they recognize wisdom in others? If we are to retake our nation from the fools who now run it, we as a nation must broaden our experience and wisdom. Lots of knowledge about a piece of the puzzle may qualify a person to earn a good living. It takes more to be qualified to select wise leaders. If our leaders destroy us, What will become of our ability to earn money?

Perhaps our leaders are not fools. Perhaps they are very wise about getting elected and care little about anything else. They claim to care, which means that if they are not fools they are liars and frauds. Fool or fraud, Which do you prefer? Which is the most dangerous? I might add (In fact, I will add) neither major party has a monopoly on fools or frauds.

The only way we will get wise leaders is to create an environment where the wise can be elected. Those who would make the best leaders rarely even run for office. They see how the superficial candidates with the glitzy campaigns usually win. Until voters find the wisdom to elect wise candidates, we will continue being governed mostly by fools and frauds.

THE EVIL OF TAXES

I visited a store in the Kalamazoo area. By today's standards it wasn't a big store. I thought about the country general store my parents owned half a century ago, where they sold groceries, meat, a little hardware, boots, shoes and some clothing. Gas pumps stood out front, while there were chicken feed and a vinegar barrel in the back room. Yet, my parents general store would fill only a corner of the store where I now stood. Could you supply your needs in this big, new store? Not unless you were a dog. The shelves were laden with rawhide bones, doggie delights, harnesses, collars, and even little sweaters, to name only a few of the available items. Those shelves extended as far as the eye could see.

The evil of taxes is not that we must pay them. If tax money was simply taken from us and thrown away, taxes would be but a minor threat to our well being. The evil of taxes is that government gets to spend them. Every tax dollar we pay to big government is another nail in freedom's coffin.

The store contained not one item intended for human consumption or use. Yet, a steady line of customers flowed in and out the doors with their "masters" following on the ends of leashes. The more trusting customers let their "masters" go to the store by themselves to procure the necessities for life in pet land. In my parents' store, a few bags of dog food set in a corner. That corner has now expanded to an entire store.

This was only one store. We now have stores devoted solely to toys, video tapes, exercise equipment, electronic entertainment, water beds, or innumerable other items that are by no stretch of the imagination necessities. Even what we buy as groceries goes far beyond the realm of necessity. Only a small percentage of our nation's income is spent on true necessities. Most of what we call "needs" are nothing more than "wants."

A recent survey shows that 60 percent of Americans feel they can't earn enough money to pay for the life style they desire. So what? I suggest that this only means 39.9 percent are lying. Human nature is such that it is impossible for normal people not to want something.

Americans are fed up with taxes, so I've heard. That's fine. It's ridiculous to say that Americans simply can't afford to pay more taxes. Our taxes could be doubled and we would still have far more left over than the citizens of 100 years ago would have had with no taxes. If paying more taxes would solve our problems, we couldn't afford to not pay the increase.

The evil of taxes is not that we must pay them. If tax money were simply taken from us and thrown away, taxes would be but a minor threat to our well being. The evil of taxes is that government gets to spend them. Every tax dollar we pay to big government is another nail in freedom's coffin. Individual freedom is inversely proportional to the tax dollars government gets to spend.

Rather than solving problems, government is rapidly destroying the will and ability of the citizens to deal with the problems of life.

The government bureaucracy operates on the assumption that we citizens are too dumb to be allowed to spend our own money or direct our own lives. Every tax dollar spent on domestic matters weaves another strand into the web of bureaucratic regulations that suffocates and enslaves us. Big government has totally failed to solve any of the problems it has attacked with its bureaucratic legions. Yet, in the process this government placed fetters on almost every aspect of our lives. We live in regulatory cages that grow ever smaller. Rather than solving problems, government is rapidly destroying the will and ability of the citizens to deal with the problems of life.

The politicians who would reform the government bureaucracy are no-better than those who would keep it the way it is. A reformed bureaucracy is still a bureaucracy. Why should we cheer for more efficiency on the part of our destroyer? Someone said, "Be thankful that you don't get all the government you pay for." Another way of putting it is, "Gridlock is the only friend we have in Washington." When government wastes your tax dollars, be thankful, as you would be if some other enemy fired his ammunition and missed. Until a solid majority of Americans demand the dismantling of the federal bureaucracy, the menace will continue to grow.

When government wastes your tax dollars, be thankful, as you would be if some other enemy fired his ammunition and missed.

To save America, we must vote for the candidates who promise the least. Then we must hold their feet to the fire to see to it that they deliver. Perhaps it would be more accurate to say, "We must see to it that they don't deliver." What we demand of our elected officials is far more important than who we elect. Once elected, most will march to the drum of whichever parade they believe is headed toward reelection. New candidates will also join that parade. That should be obvious this year [1994] as we see Democrats fleeing from their President and trying to sound conservative.

The direction the candidates of either party march after the election, depends on which way they believe the voters' parade is headed. Sure, politicians will try to steer the parade. When politicians are unable to dictate the direction of march, they don't refuse to march. What the voters do on election day is important. What the voters do the other 364 days of the year is vital. When the voters are away, the politicians will play – with our dollars and lives.

BATTERED

One of the many maladies that gained attention in the late 20th century is the "battered spouse syndrome." Some women will put up with just about anything from their husbands or boyfriends. These women may complain from time to time. The one thing most won't do is get out. Occasionally after a particularly abusive incident a woman makes a break. It usually takes no more than an empty promise to get her back for another round of degradation and abuse. As often as not she returns and grovels for her abuser to forgive her and take her back. The abused often deny the abuse and defend the abuser.

Why does it happen? Some people have so little self confidence and self respect that they are certain they can't survive on their own. An abusive relationship is better than none, so they believe. It is unfortunate that millions of people, many but not all of whom are women, participate so willingly in their own degradation and abuse.

During the nineties the citizens of this nation may well be experiencing the last chance to regain mastery of their lives and reassert their freedom. If the abused voters fail to break the bondage of the parasitic big government in the next few years, the bonds may well grow too strong to be challenged.

If this capitulation to abuse affected only a few million spouses it would be tragic, but not a disaster, except for those who were abused. There is a companion malady that reaches out its tentacles to torment nearly every citizen of our land. It could be labeled the "abused voter syndrome."

Tens of millions of voters have been so abused and degraded for so long by the bureaucracies of big government that they accept it as the only way of life. Occasionally these abused voters snarl at their tormenters. In the end they fall back into subservience. They fear that they could not survive without their abuser. Each time the abused rise up and then reacknowledge the rule of their heinous masters, the bondage grows stronger. This reduces the chances they will ever escape.

Those of us who do not willingly accept the abuses of government suffer with the willing victims. For so long as the willingly abused continue to grovel at the feet of the hydra headed bureaucracy, and cast a majority vote for sustaining the monster, we will all suffer the consequences. Many who vote to sustain the monster deny they are abused. We should take these denials no more seriously than those that flow from the bruised lips on the battered face of a beaten wife.

The aroused anger of the abused voters may last through an election. Cheer not if it does. If these voters again grovel before the beast two years later, all that is accomplished is that the chains of bondage are forged ever tighter. During the nineties we may well be experiencing the last chance to regain mastery of our lives and reassert our freedom. If the abused voters fail to break the bondage of the parasitic big government in the next few years, the bonds may grow too strong to be challenged.

In the struggle for freedom we are in the fourth quarter trailing by at least a couple of touchdowns. In the 1994 election we gained a first down on our own 15-yard line. That's not much. It's far better than losing the ball. It kept the drive alive and gave us a chance to further the advance of individual responsibility and freedom in 1996.

Until millions more voters stand up to their abusers and accept the risks and opportunities of living free, we cannot end the exploitation and abuse by an ineffective and worse than useless federal government. Some functions of the feds are necessary. The federal bureaucracy must be more than decimated before we come close to pruning it back to its beneficial functions. Until those now willingly abused grab a saw and pruning shears and join in the trimming, there is little reason for hope.

RALPH

On a long past summer day a rather unpretentious looking black and white cat took up residence among the barn cats already inhabiting my realm. For reasons I do not recall, he came to be known as Ralph. Nothing distinguished Ralph from his piers. He passed his time doing the usual cat things, like eating and laying motionless in the sun for hours at a time.

Alas, one day I visited the granary only to discover that a colony of mice had staked a claim in the oat bin. Not being much at chasing mice, further having no desire to grab a live mouse even if I got the chance, I sought aid. There was Ralph, well rested and ready from a morning spent laying in the sun. The big question was, Ready for what? Tomcats are not famous for their hunting powers. When there is only one option, it makes the decision much easier. I picked up the listless Ralph and marched back to the granary. What did I have to lose?

Actually Spot is his last name. It is of Shakespearian origin. When he was riding my toe out the back door for the umpteenth time, I christened him, "Out, Out Damn Spot."

I was not prepared for what happened when I unceremoniously dropped Ralph among the fleeing rodents. In one of those other nine lives, Ralph must have been a Green Beret. Before I could blink, two wiggling tales protruded from Ralph's mouth. He was in hot pursuit of the third. Then life grew complicated. Everything has a capacity limit, including Ralph's mouth. When he captured number four, number three escaped – Or was it number one? Who cares?

Ralph pursued his game of musical mice with reckless abandon. Shortly, the mice succeeded in immobilizing Ralph. With three in his mouth and one under each front paw, Ralph could only roll his eyes in frustration while his captives' companions scurried for cover. Sensing that the mighty warrior was on the verge of a nervous breakdown, I came to his aid. I carried him, with his captives, to a quiet area distant from the battlefront where he dealt efficiently with his prisoners. Moments later Ralph, unburdened of the prizes from the first battle, made his next sortie to the front. When the dust finally settled, the granary was again safe for oats. A somewhat bloated Ralph returned to his place in the sun.

Whenever the rodents penetrated the granary's defenses, Ralph answered the call to duty. He served with such vigor and valor that he earned a peerage. Even in Camelot, time and tide waiteth for no man, or cat. Sir Ralph went to his reward.

In recent times another black and white cat ventured into the realm. Could he be the reincarnation of Sir Ralph? The new arrival fancied himself a house cat rather than a barn cat. Thus, he earned his name, Spot. Actually Spot was his last name. It is of Shakespearian origin. While he was riding my toe out the back door for the umpteenth time, I christened him, "Out, Out Damn Spot." You may recall from a previous chapter, I have no problem with two middle names.

Again the rodents invaded the granary. It was time to put Spot to the test. Sir Ralph would have considered it an insult rather than a test. One puny mouse cringed in the bottom on an empty feed barrel. I dropped Spot into the barrel, somewhat like a Green Beret floating down from the sky. That was where the similarity ended. The adversaries stood nose to nose. Spot looked at the mouse. The mouse looked at Spot. The mouse retreated to the other side of the small barrel, behind Spot. In his panic, the mouse climbed up Spot's tail. That was enough for Spot. It was time for action. He jumped out of the barrel. I should have named him Garfield.

I could but wonder if up in that great granary in the sky, Sir Ralph covered his face in embarrassment for the level to which his species had fallen.

Does It Matter Who Controls the TV?

Again we hear technology will save us. Our saviors will implant the V-chip in television sets. It will block out objectionable programs on command. Then parents can stroll blissfully through life assured that the minds of their little darlings are safe from the ravages of violence, perversion and whatever other mental tripe the lords of the electronic media dispense.

Lurking dangers have threatened the bodies and minds of children throughout history. As our environment changed, the nature of the threats changed. As always, it's the job of parents to first protect the children, and then to teach the children to protect themselves.

Who will make the decisions about what is filtered out? – The same people who produce the garbage in the first place. Isn't that comforting? Parents who rely on such devices will still be letting the TV barons mold their children. Only the methods will have changed. Might I be so cynical as to suggest, these devices aren't even intended for protecting children from garbage? Rather, their purpose is to protect TV executives, producers and sponsors from parents and Congress.

Parents who need an electronic device to control their children's TV viewing have already lost the game. They are only waiting for the last out so they can crawl back into the dugout and wonder why they lost. Raising children requires the presence and involvement of parents. There must be a relationship. Establishment of discipline is essential. If the children do as they please whenever the parents aren't looking, the parents have already failed. No electronic device can compensate for that. Sure, even the best disciplined of children will sneak a look at the forbidden programs now and then. An occasional look is no big deal. It's the steady diet that does the damage.

What would happen if TV devoted equal time to carefully crafted spot messages showing the black side of alcohol?

Some of the most devastating impact of television comes from commercials. Commercials pop up all the time in all programming. One example. Can any rational person deny that growing up on beer commercials contributes significantly to the teenage alcohol problem? The theme of most beer commercials is, those who drink beer have wonderful times, and drinking is indispensable to having a wonderful time. Can you blame the kids for being eager to join in all that fun? What would happen if TV devoted equal time to carefully crafted messages showing the black side of alcohol?

Looking to electronic devices to protect children from the negative effects of television is as futile as looking to gun control to protect us from crime. Both ideas have some appeal, until you dig beneath the surface to discover that they simply don't work.

Lurking dangers have threatened the bodies and minds of children throughout history. As our environment changed, the nature of the threats changed. As always, it's the job of parents to first protect the children, then to teach the children to protect themselves. That's what growing up is all about. Even if the electronic device provides protection, it will do nothing to perform the more important task of teaching children to protect themselves.

We keep matches away from small children. A teenager who can't be trusted with matches is a serious problem. It's impossible to keep matches from a thirteen year old. The same is true of television. Parents must teach responsible use of television, and teach about its dangers. The real danger of relying on an electronic device is that it lulls parents into a false sense of security. They then do an even poorer job of preparing children to live in a world dominated by television and its many negative messages.

As a society, we have downgraded the importance of raising children, and reduced the amount of effort we are willing to put into the job. The lame excuse that there isn't enough time, is nothing more than a self serving copout. For the most part, parents don't have time because they are so busy earning and spending money for their own benefit and amusement. It's worse than silly to even suggest that our ancestors, who worked long hours seven days a week, had more time to raise children than we do.

Each generation starts with a blank slate. Any generation will have only as much civilization as its parents and other elders ingrain into it. Until we reestablish the importance of raising children, and invest the required effort, our civilization will continue to decay. Crime bills, health care, welfare reform, and even our efforts to improve the schools, are all futile if we don't get serious about raising children. This is the one indispensable task for the survival of civilization. Unless we change, and change quickly, the epithet for our civilization may well be, "They let the children raise themselves."

THE WAY

Rarely do we get up close glimpses of history such as those in *The Private Life of Chairman Mao,* written by Dr. Li Zhisui, Mao's personal physician for over 20 years. So far I have only read excerpts from this book. These excerpts contained more worthwhile reading than most entire books.

In the late 1950s during the Great Leap Forward, Dr. Li traveled through the countryside with Mao on his private train. Dr. Li wrote of his trip. "All the able-bodied men, the farmers of China, had been taken away to tend backyard steel furnaces, another innovation of the Great Leap. We could see them feeding household implements into the furnaces and transforming them into rough ingots of steel. I don't know where the idea of the backyard steel furnaces came from. But the logic was: Why spend millions building modern steel plants when steel could be produced for almost nothing in courtyards and fields? Furnaces dotted the landscape as far as the eye could see."

Anytime a person's future depends on pleasing the boss, pleasing the boss becomes the mission. If pleasing the boss means fooling the boss, So what?

"The claims for both grain and steel production were astounding. Mao's earlier skepticism had vanished. I was infected, too. . . ."

Later, Dr. Li learned the entire scene was a huge, multi-act opera, performed for Mao. Local party secretaries ordered the furnaces constructed for three miles on either side of the route. One province party chief ordered peasants to bring rice plants from other areas and plant them along the railroad to give the impression of an abundant crop. Production figures were false. The product from backyard furnaces was useless. The steel shown to Chairman Mao came from real steel mills.

The book would be interesting if it merely revealed some insight into happenings in China a few decades ago. As is so often the case with history, the message is far more significant than that.

The show staged for Chairman Mao reminds me of my former employer, a large corporation in the heart of Michigan. The managers and underlings treated their chairman, and other members of upper management, the same way Mao's underlings treated him. They would half do a job to get it done by the time demanded by the chairman. They would pretend things were done that weren't, hoping to fill the gaps before the truth came out. They staged a show for the boss.

I told a lower level manager that a project couldn't be completed to meet the chairman's schedule. He responded, "I'm not going to be the one to disappoint the chairman."

People are the same around the world. Capitalists in America readily fall into the same traps as Communists in China. When a person's success depends on pleasing the boss, pleasing the boss becomes the mission. If pleasing the boss means fooling the boss, So what? The mere fact that attempts to please the boss are detrimental to the enterprise, makes no difference in business or government.

To motivate people to make an enterprise succeed, their success must be measured by the success of the enterprise, not by the boss's pleasure. A farmer isn't going to fake spring planting to make his fields look pretty. He knows, the fall harvest, not the neighbors' opinions of the fields, measure his success.

Here in lies the simple, yet often obscured, truths as to why all large enterprises eventually crumble and fall. The boss at the top doesn't know what is happening. The success of his underlings depends on pleasing the boss. Underlings compete in a great game of show and tell to win favor with the boss. The best actors, salesmen and prevaricators, not the best thinkers and doers, prevail. The whole operation becomes a farce.

An enterprise facing competition is headed for big trouble when it wastes its energies pleasing the boss. A government without competition can stumble on until the whole nation collapses. Thus, though big businesses are infected by this disease, their infections are usually minor compared to what happens in big government.

Drop a rock and it will fall. Laws of human behavior guarantee the failure of any government that tries to manage the economy and life of an entire nation. These laws are as ridged and certain as the law of gravity. The only difference is, the length of time required for the fall cannot be predicted with the same degree of certainty.

IS PRAYER CONSTITUTIONAL?

Again the rumblings from Washington include a "prayer amendment." What inspired this turn toward religion? Is it more than the beneficiaries of the votes of the faithful trying to pay a political debt? Is there a true ground swell of enthusiasm for making prayer a part of our public schools? What would be the ground rules for prayer? Who would write, interpret and enforce the rules of prayer? Do we really want to assign these tasks and powers to any government?

I fail to see how religious beliefs and the practice of religion is going to benefit from pushing a prayer amendment, whether it's adopted or not. It would seem that at best the amendment would be a statement of principal that wouldn't change the law. At worst it would be the opening shot in a war to establish politically correct prayers.

Is the proposed amendment an over reaction to the over zealous acts of a few anti prayer extremists? I heard that a school principal disciplined a student for bowing his head in the cafeteria to pray before his meal. A traffic cop could give you a ticket for failing to stop at a green light. That doesn't mean the traffic code should be amended. The traffic cop should be retrained or dismissed.

I have not detected any consensus about what a prayer amendment should allow. It's easy to say, "Let's adopt a prayer amendment." It is not quite so easy to write one that pleases many people. If we sanction audible prayer at public school functions – Who shall pray? To Whom? For what? If we allow one, Must we allow all? If we do not allow all – Who chooses the prayers? If a government official or other public employee chooses – Is this a dangerous mixing of church and state? Our recent history suggests that judges will end up writing the prayers. Who's ready to vote for that "solution?" Will anyone, whether majority or minority, be satisfied, if he finds the prayer offensive?

Is Prayer Constitutional?

Fifty years ago prayer and religious observances in public schools were commonplace. Fifty years ago we lived in a different world. The people in many of our communities shared common religious views. This, coupled with the fact small minorities usually were sufficiently intimidated so as not to complain very loudly or very long, made the system work fairly well.

Today the opposite conditions exist. Small minorities actively seek to stomp out anything which hints of religion, even if no one in the community is offended. These agitators come from outside the community to stir up trouble. A pox upon their houses. The troublemakers aside, in most communities today a prayer sufficiently sanitized to offend no one, won't please anyone either.

> **It's unrealistic to expect students to check their religion at the door like guns at a wild west saloon. Schools should accommodate religious expression, and religious clubs and organizations as much as other forms of expression and groups.**

Perhaps the backers of the prayer amendment have in mind only to reaffirm the right to pray silently wherever you choose. Such an amendment would do little, if anything, to change the law as it exists today. Perhaps it would serve to restrain a few over zealous prayer bashers who now go beyond the law. Should we undertake to amend our Constitution only to underline what the law really is now? The lawless will still disregard the law. If the amendment is promoted and fails – Won't some interpret this as meaning there is no right to pray silently?

How will religious beliefs and the practice of religion benefit from pushing a prayer amendment, whether it's adopted or not? At best the amendment would be a statement of principle that wouldn't change the law. At worst it would be the opening shot in a war to establish politically correct prayers. Might this war produce far more losers than winners? How many of our scarce resources of time and money should we consume in the battle? Might the energy be far better spent assuring that public schools are neutral on religious practice – in other words, not hostile to it – rather than trying to use schools for the open practice of religion?

It's unrealistic to expect students to check their religion at the door like guns at a wild west saloon. Schools should accommodate religious expression, clubs and organizations as much as other forms of expression and groups. Students should have as much right to express their religious beliefs as their political beliefs. No one should expect the right to make his political or religious practices part of the official school program. We don't need a constitutional amendment to allow religious expression in public schools. We need only to bring school practice into line with the law as it exists. Until we pursue this route, Might it be well to put the constitutional amendment on hold?

Might supporters of a prayer amendment ask themselves – How would they feel about school prayer if they were in a minority that would never have any voice in the prayers to be offered? What if they found those prayers offensive to their beliefs? Remember, any of us could be in that minority someday.

THE JOYS OF JANUARY

Ahh, January. When else can I awake to a foggy, slushy, sloppy morning and feel good because the weather has improved? When I ventured out into the slop, I discovered I was not the first to arrive. A pile of freshly excavated soil bore witness that the woodchucks living beneath my barn had awakened exactly three weeks before Ground Hogs' Day.

Perhaps the two deer that escaped weren't assailants at all. Maybe the one that nailed the car was a deer nerd, and the others set him up.

If I were a bit more optimistic, it would have surprised me that woodchucks still lived beneath my barn. Last year I captured fourteen or more in that barn and the lawn in front of it. Varmint activity rose to such a high level that I bought a second live trap in a desperate attempt to salvage a few strawberries and a little sweet corn.

My use of live traps is not motivated by compassion for the varmints. The live trap lets me choose what lives and what dies. Trapping in my own barn and garden occasionally results in the capture of a cat or the like. If I used a kill trap, it would be a case of, "Kill them all and let God sort them out." I did take Joycelyn Elders advice to use safer bullets. By using .22 shorts, instead of long rifles, the bullet remains in the varmint. This is much safer for the trap. Varmint safety isn't one of my concerns.

Varmints of another kind are still out in force. Last Sunday evening shortly after night fall, I ventured to the window. I started to close the shades to block out the cold and gloom. Behold, three sets of taillights glowed from the road just beyond my garden. Three sets of moving taillights on my road are unlikely. Three sets of stationary taillights are all but unheard of. This may well have been the first time in twenty years.

Shortly, one of my neighbors and a stranger appeared at my door. The increasingly hostile deer of the neighborhood had assaulted another car. My investigation of the scene revealed that three deer used my front lawn as their staging area for the attack.

They apparently grew impatient waiting for a victim. They thoroughly trampled the snow in my lawn while they paced about anxiously. Previous kamikaze assaults from my lawn were the work of lone assailants. The gang attack is disquieting.

Two of the assailants were unsuccessful and loped off into the woods. The third scored a direct hit incurring three broken legs. He still made it fifty feet into the field where I administered the *coup de grâce* with an ax. Perhaps the two deer that escaped weren't assailants at all. Maybe the one that nailed the car was a deer nerd, and the others set him up.

To all you anti hunters who may read this (both of you), Is this really a better way for deer to die, rather than being shot by a hunter? Remember, this scene was repeated more than 55,000 times last year, just in Michigan. I recall years when hunters took only about 60,000 deer. With the unreported deer hits, chances are that drivers now take more deer than the hunters of a few years ago.

The next morning I looked out my window to find my entire lawn and garden crisscrossed with deer trails. There wasn't a ten foot square of untouched snow. Deer have frequented my lawn and garden every since I moved here. In recent years the activity has been on the increase. Until now though, three or so trails at a time were the limit.

The DNR says the deer herd is smaller. Should we assume that the DNR is the one government agency that won't lie to us? For so long as I can remember the DNR has advocated more deer. The ever increasing problems caused by deer have diminished the popularity of increasing the size of the herd. For the DNR to get away with further increases in the herd, the easiest way might be to claim the herd is shrinking. The DNR could expand the herd while pretending they are only trying to maintain it.

A large deer herd will accomplish several things. One of those is, more and more people will come to view deer as nothing but big rodents with antlers. Sooner or later the public outcry will mandate decimation of the herd. Thus the advocates of a large deer herd will achieve a much smaller herd. This is another illustration of the old adage – In politics one usually achieves the opposite of what he (or she) attempts. This could be called the Hillary Clinton syndrome.

THE LION ROARS

The mountain lion controversy flares again in California. That's one of the best reasons for having California. We have Michigan's Upper Peninsula for storage of snow. We have California for the storage of ridiculous controversies. Both sometimes overflow.

Today's mountain lion controversy is the inevitable result of the first round in the 1970s. By referendum, Californians banned mountain lion hunting. At least it wasn't imposed on them by Washington or any other outsiders. Anyone who thinks all foolishness is in Washington, please take note. What about the 48 percent of Californians who voted "no" on the hunting ban? What about those who were too young to vote? Do they enjoy playing mountain lion roulette more because neighbors, rather than outsiders, signed them up for the game?

The hunting ban is one example of a law that worked. The mountain lion population exploded. So what? After all we are talking California, the home of Disney. Didn't Disney create mountain lions, and most other animals, in his own image?

> ***U.S. News*** **reported that "A Declaration on Great Apes" says, chimps, gorillas and orangutans are an oppressed group with the right not to be imprisoned without due process. Would it be politically correct to grant lesser rights to mountain lions? Should execution require less due process than imprisonment?**

The nation took note when one of those protected lions ate a somewhat less protected jogger. This incident did little to cool the ardor of the mountain lion zealots. So what if the lion ate a person? We still have more people than lions. To suggest it is worse for a lion to kill a person, than for a person to kill a lion, would be to prefer one species over another. This would not be politically correct.

Californians are not quite so politically correct as they like to believe. They summarily executed the killer lion. It wasn't granted the same ten-year waiting period given human killers before execution. Even in California utopia is still down the road a piece.

One mountain lion enthusiast said we should be willing to accept a couple of deaths per decade as the price for having lots of mountain lions. Well, that second death has now occurred. Does this mean that with the mountains lions having taken their decennial quota of humans, Californians are safe until the end of the decade? I'll leave the debate on that to the Californians.

I will take the liberty of suggesting some questions they might consider. What if no one told the lions about the quota? What if lions are like human hunters who sometimes poach? What punishment should be imposed on a lion who poaches? What penalties should be imposed on humans that take the law into their own hands and retaliate against poaching lions, rather than waiting for the due process of law?

U.S. News reported that "A Declaration on Great Apes" says, chimps, gorillas and orangutans are an oppressed group with the right not to be imprisoned without due process. Would it be politically correct to grant lesser rights to mountain lions? Should execution require less due process than imprisonment?

In the name of humanity, we must send psychiatrists to California to deal with the stress and nervous breakdowns that are inevitable as the Golden Staters wrestle with the great issues of our day.

National Review reports that a mountain lion attacked a five year old girl leaving her with permanent brain damage, partially paralyzed and with only one eye. "Still some people who called the Fish and Game Department were adamant that the lion not be killed. 'It was terrible what happened to . . . ,' one caller said. 'But you can always replace a human. . . . , You can't always replace a mountain lion.'" On reading the above, it occurred to me there are some humans we would be well advised not to replace.

Do those of us who live in the other 49 states have any reason to take comfort because this epidemic of the terminal sillies ravages California rather than our domains? Are our fellow citizens cut from more rational cloth? Don't bet on it. In Michigan we owe our lack of a mountain lion controversy far more to the small number of mountain lions (roughly zero) than to the shortage of wacky people. We can feel fortunate that our predators, such as coyotes and foxes, usually confine themselves to eating members of the non human species. In Michigan even the animals aren't politically correct. Perhaps some day we can be as enlightened as the Californians.

Before Michiganians rush to stake a claim to being more enlightened and rational than Californians, it might be well to note that Michigan's deer kill and maim far more humans each year than do California's mountain lions. Granted, the deer don't eat people. Is the human killed by a deer any less dead because the deer acted without malice?

DOES IT WORK?

A television news report featured a system using private rehabilitation instead of criminal prosecution for misdemeanors. The defendants were given the opportunity to choose the private program rather than being tried for a criminal offense. The defendant must pay the cost of the program and successfully complete it.

According to the report, this program shows impressive results. Only about 10 percent of those completing the program were arrested again. If we are to solve the problem of deterioration of the quality of life in our land, we must have vision and try new approaches. I would not attempt to judge the success or failure of private rehabilitation (or anything else) based only on a TV report. Certainly though, the idea should have a chance to prove itself.

Government today is calling the plays for businesses, rather than making the rules. Wrong minded government intervention screws up the free enterprise system.

The most disquieting parts of the report were the statements of a critic of the program. The critic observed that because this program was conducted by private business, it was being done to make money. This can come as a surprise only to the truly ignorant. It is not a particularly well kept secret that the main mission of any business is to earn money. If earning money is evil, then by definition all successful businesses are evil.

The critic further observed that the business made millions of dollars. The critic's message was – The business sought profit; therefore, the program was bad. I don't know if the estimate of the profit level was correct. So what if it was? Who cares? More to the point, Who should care? When a business make millions of dollars – Does that automatically render the product bad? The gross ignorance and prejudice displayed by this critic are part of the burdens dragging America down. I fear that millions of our fellow citizens share the view that something is inherently evil about any enterprise that sets out to earn money. The enterprise that seeks profits is presumed evil. By achieving profits the enterprise proves itself evil.

When a business providing a good product or service earns solid profits, free enterprise is working. Profits are the reward a business should receive for doing a good job. How hard would you work if you didn't receive any reward? – such as a pay check. The role of government should not be to limit or obliterate profits. Competition in the marketplace should be the only control of profits. The only role of government should be to establish the rules of the game so that a job well done can earn profits. The rules should not allow monopolies, fraud, intimidation, and deceit to be profit makers.

For athletic contests the governing body establishes rules allowing for fair competition – allowing talent and ability to prevail, rather than rewarding manipulation and intimidation. The commissioner of the NFL doesn't intervene to adjust the scores of the games, or to make sure no one wins. He would defeat the purpose of the game if he did. If there are problems in the games, the league adjusts the rules. It doesn't send in league agents to call the plays for all the teams.

Government today is calling the plays for businesses, rather than making the rules. Wrong minded government intervention screws up the free enterprise system. Then government self righteously declares that only more government manipulation will solve the problems government created. If we are to straighten out the mess, government must get back to making reasonable rules, rather than calling the plays. As a first step we must overcome the type of economic ignorance shown by the critic of the private rehabilitation program.

The Miracle of Increasing The Minimum Wage

Few things please politicians more than helping people, especially when elections loom ahead. One of those helpful things politicians do from time to time is raise the minimum wage. How helpful is it to raise the minimum wage? Certainly, those working for minimum wage will find the pay increases helpful. Skeptics suggest that those who end up with no job at all because of the higher minimum wage, may not have been helped quite so much. Is a little unemployment too high a price to pay for higher wages for millions?

The people who take the biggest hits from minimum wage increases are those with fixed incomes, such as pensions. The ripples from the minimum wage increase don't reach their ponds. The inflation does.

What happens when Congress raises the minimum wage? Almost immediately those workers earning slightly more than minimum wage get pay increases too. The splash from the minimum wage increase sends ripples up the wage chain increasing the pay checks for tens of millions of workers.

Before the waves die out, nearly every worker shares in the higher pay. If this didn't happen, successive minimum wage increases would compress the wage scale until everyone worked for minimum wage. How can an act of Congress increasing the pay for nearly all Americans be a bad thing?

Who pays the higher wages? They aren't paid by the congressmen who vote for the minimum wage increase, or by the President who signs the new law. An increase in the minimum wage doesn't increase productivity. The employees continue doing the exact same job for higher wages.

Where do the employers get the money to pay higher wages? If the employer is government, it raises taxes. If the employer is a business it raises prices. Those who can't raise taxes or prices, hire less workers. That's why some jobs are lost.

THE MIRACLE OF THE MINIMUM WAGE INCREASE

When prices go up, we call it inflation. By the time the ripples from the minimum wage increase settle down, the price of almost everything has gone up. Those working for minimum wage, along with everyone else, are no-better off than before the increase in the minimum wage. Many people are worse off.

The people who take the biggest hits from minimum wage increases are those with fixed incomes, such as pensions. The ripples from the minimum wage increase don't reach their ponds. The inflation does. There is a silver lining. When the retired take minimum wage jobs to supplement their shrunken pensions, they will get that higher wage.

The old adage, "Enjoy it while it lasts," certainly applies to increases in the minimum wage. Those who benefit from the first wave of wage increases actually experience a brief increase in purchasing power. As the wave spreads across the economic pond, the benefits fade away.

Who benefits from increases in the minimum wage? Only the politicians. At first glance an increase in the minimum wage can look pretty good. Millions of voters never take the second glance. Politicians gain votes by pawning off an increase in the minimum wage as a good thing. As the TV adds say, "Wait, there is even more." Due to the graduated income tax, a higher percentage of almost everybody's wages is gobbled up by taxes after the pay increases. Raising the minimum wage is a devious way to increase taxes, without ever mentioning the dreaded T word.

This may explain why President Clinton's most passionate plea in his 1995 State of the Union Address was for an increase in the minimum wage. An open plea for a tax increase on everyone would not have been enthusiastically received. I said politicians were helpful people. For some reason they always work the hardest when they are helping themselves.

HAVE WE BURIED INFLATION?

Inflation is dead. At last we can forget it. Is this really true? After reading of his demise, Mark Twain told the world the report of his death was greatly exaggerated. Are the reports of the end to inflation exaggerated? Does the beast still roam among us?

In 1971 President Nixon imposed wage and price controls in an attempt to stop inflation. He was driven to this drastic action because the inflation rate rose above 4 percent. A little over ten years later inflation dropped below 5 percent. Everyone was relieved and pleased that inflation was conquered.

Considering that profits aren't increasing, there are only two causes of inflation. Increased taxes and increased labor costs.

Now it's a matter of routine to see or hear news stories reporting that there isn't any inflation. If you look closely, you will discover that all these articles report an annual inflation rate no lower than the 3 percent range.

In 1971 four percent plus was unacceptable runaway inflation. Now 3 percent plus is no inflation at all. What a difference one little percent makes. Should we be so unconcerned about the beast eating our dollars just because he takes smaller bites?

What is inflation? Simply put, inflation is higher prices. Wages may go up too. When prices go up we have inflation, whether or not wages go up. Why do prices go up? There are only three reasons. It cost $25.00 to make a pair of shoes. The shoes sold for $30.00. Then the price rose to $35.00. One of three things happened. The cost of making the shoes may have gone up. Taxes may have gone up. The other possibility is that the shoemaker increased his profit. The price increase might be part cost increase, part taxes (another form of cost increase), and part profit. There is absolutely no other way for prices to go up.

Business profits over the years have not gone up. This means there is only one cause of inflation. The cost of producing the goods and services we buy (including taxes) has increased.

Have We Buried Inflation?

Does inflation matter? Interest on savings accounts is about 2.5 percent. Inflation is over 3 percent. Put $100 in the bank. At the end of the year you have $2.50 interest on which you may pay $.75 income tax. You now have $101.75. Are you a winner? Don't forget that it now takes over $103.00 to buy what you could have bought with the $100 at the beginning of the year. You lost $1.25.

Big deal. What is $1.25? Keep doing this for ten years. Your $100.00 will shrink to $88.50. Thirty years later only $69.00 would remain. The same erosion eats fixed pensions. The pensions draw no interest, thus accelerating the erosion. It only takes 12 years of 3 percent inflation to shrink $100 of pension to $69. These are only a couple of the ways that inflation, even "low" inflation, gets its hands in our pockets. Would we so willingly accept a blackmailer or robber who called each year to take the same bite?

What causes those cost increases that fuel inflation? The cost of the goods and services we buy is the cost of labor, materials, equipment, supplies and taxes. The cost of material, equipment and supplies is the cost of the labor, materials, equipment, supplies and taxes to produce them. Follow this chain to its beginning and you will discover that all the cost of everything is labor and taxes. Small amounts may be paid to landowners as royalties for the mining of raw materials. These payments are in reality just another form of tax, or compensation for the labor of securing the land.

Considering that profits aren't increasing, there are only two causes of inflation. Increased taxes and increased labor costs. Yes, tax increases on business do cause inflation. Inflation or lower profits are the only alternatives. Keep cutting profits and the business will go bankrupt. This causes unemployment and a whole lot of other problems. (You may have heard that profits are going up. These increases are inadequate to make up for the decreases that slipped by unnoticed.)

The only other cause of inflation is increased labor costs. There are two ways to increase labor costs. One is to cut productivity. If each worker produces less, the business has to hire more employees to do the same work. More employees mean more wages. More wages mean higher prices. For the most part productivity hasn't gone down since inflation got rolling in the sixties. Loss of productivity is not the cause of inflation.

What is left? Only one thing. Higher wages. Increase wages without increasing productivity, and businesses have no choice. Prices must go up. Higher prices are inflation. Trying to beat inflation with higher wages is like a dog chasing its tail. You never catch up. When one worker gets a pay increase big enough to beat inflation, he takes it out of the pockets of those who didn't get the same pay increases.

Why do we allow these useless and damaging pay increases to continue? Are business and government leaders so ignorant that they don't understand what is happening? No. Most of them know exactly what is happening. Everyone want's a pay increase. It doesn't matter that the employee may be doing the exact same job as last year. The annual pay raise has become such an American tradition that it's almost impossible to refuse it. It is often easier for business to give in to these inflationary pay increases than to fight them. As long as everyone gets hit with the same increases, it doesn't change the competitive position of the businesses.

Remember though, with growing world trade "everyone" includes the whole world. That's why business took a harder stand against inflationary pay increases in the eighties. Inflation in this country was killing us in competition with foreign businesses. As long as inflation doesn't greatly hurt the market for a business's products, the business will go along with inflationary wage increases demanded by the employees.

As long as millions of employees get pay increases without corresponding increases in productivity, inflation must continue. The size of the pay increases will determine the rate of inflation. Those pay increases aren't paid by the employers. They are paid by the employees themselves, and their friends, and neighbors. The only winners are the ones whose increases exceed the rate of inflation. They gain their increases by picking the pockets of people on pensions, people with savings accounts, and anyone else who doesn't get a pay increase just because prices go up. For every winner there is a loser. Would it not be far better to end those inflationary pay increases? Let everyone keep what they have and quit picking each others pockets. As in all competitions the weakest lose. Those who can least afford it, pay the price for inflation.

TAXES MAY NOT LIE – BUT LIARS TAX

It's the tax cutting season. A short time ago the air waves were filled with voices equating tax increases with patriotism and sainthood. Now to even voice opposition to tax cuts is political suicide. The battles are over where to cut and how much – most of all over which politicians will get the credit.

What elevated the tax cut from sinner to saint? If those of us in the grandstands at the tax cut three ring circus are to truly appreciate the shows staged for our benefit, we must understand a few things about the rules of the game. The rule book for the taxation game bears the rather forbidding title of "Economics." There is an all too common belief that economics is the province of ivory tower professionals. Economics, like Einstein's Theory, is not for the common man.

> **One unalterable principal of taxation is that <u>businesses do not pay taxes</u> – never have, never will. Those who would have us believe that taxes on businesses are substitutes for taxes on us are frauds and liars, or at best ignorant.**

It is understandable that we shun economics and assume economic theory is beyond us. Before college, my own economic education was limited to a few tidbits spilled into the classroom, almost by accident. Early in grade school I did learn about hidden taxes, a lesson that seems largely unlearned today.

Political careers are based on the simple assumption that any tax we don't see, we will gladly pay. It works. The politicians lie to us saying someone else is paying the taxes. If we don't see how those taxes are passed on to us, we are likely to believe the lies.

I entered college with an understanding of economics equivalent to illiteracy. A couple of economics courses forced on to me opened my eyes to the existence of economics and the importance of everyone understanding the basic principles of economics.

These basic principles are simple and easy to grasp. So why does knowledge of these principles remain so rare? It isn't reasonable to expect people to seek knowledge that they don't know exists, or that they consider beyond understanding. The result is, as a nation we live in the economics Dark Ages. On election day we choose among economic theories that we don't understand. We are as likely to vote for witchcraft as for science.

In reality, true economics is neither witchcraft nor science. Economics is a branch of psychology. It deals with human behavior regarding money, and the creation and distribution of wealth. Most of the formulae and equations of economics do not deal with immutable laws of nature. They seek to predict how human beings will react to certain changes in the economic environment.

Charting the course of future behavior in the world of money often confounds the experts. We can grasp the basic principles of economics without concerning ourselves with the difficult areas and exotic theories. It is the basic principles we must understand to detect political double speak, and to vote intelligently.

A modern factory may be complex and difficult to understand. The bedrock on which it stands is a far different matter. The same is true of the bedrock of economics. It is the foundation of economics, not the superstructure of complex theory, we all need to know.

The first of the big lies in the economics of taxation goes back to those hidden taxes my grade school teacher laid bare before me. Whenever the politicians tell us about taxes, we inevitably hear about good taxes and bad taxes. Bad taxes are the ones we pay. Good taxes are the ones someone else pays.

So, tax businesses and tax the rich. Most of us are not rich and don't own a business. Thus, we are home free. Believing this is as dangerous as believing the Internal Revenue Service is a tax shelter. One unalterable principle of taxation is, <u>businesses do not pay taxes</u> – never have, never will. Those who would have us believe that taxes on businesses are substitutes for taxes on us, are frauds and liars, or at best ignorant. Business taxes are as mythical as unicorns.

Businesses have no money with which to pay taxes. Businesses have only the money they receive from investors and customers. If a business uses its investors' money to pay taxes, the business is losing money, and will soon be a business no more. The only other sources of revenue are customers and donors. Few people, except politicians, are lining up to make donations to businesses. In order to pay taxes businesses must get the money from the customers.

Businesses don't pay taxes, they only collect taxes for government. Those taxes are added on to the prices and passed on as hidden taxes. When we buy at retail we generally have no idea how much of the purchase price is taxes. Ignorance is bliss. Paying a tax is like passing a killer on the street, we feel better if we don't recognize either villain. Every tax paid by General Motors is just as much a tax on car buyers as if the tax were added on to the final price as a sales tax.

The clearest example of the pass through of business taxes is in the utility business. When the Public Service Commission sets utility rates, it first calculates the utility's costs, which include all taxes paid by the utility, except income tax.

The Commission then adds on an amount – say 10 percent – it deems suitable for profit. Next the Commission adds enough more money for the utility to pay its income taxes. Thus, the utility will have the opportunity to earn a 10 percent profit after income taxes. The Commission sets rates designed to allow the utility to collect the total amount determined, including taxes. If 15 percent of the revenue earned by the utility is for taxes – then the customers pay 15 percent more than if there were no business taxes.

All utility taxes are calculated and added on to our utility bills. To us they are hidden sales taxes. If taxes go down, public service commissions are usually quick to reduce utility rates accordingly.

All other businesses add taxes the same way. It just isn't as open and obvious. There is a name for businesses that don't pass their tax costs on to their customers. They are called former businesses. A business that doesn't pass its tax costs on to the customers goes out of business.

Former businesses not only don't pay taxes, they don't collect them either. "Business tax" is a synonym for "hidden tax." We would be better off if there were no business taxes. If we paid all our taxes up front where we could see them, I have a feeling we would be far less tolerant of government waste.

Anyone who advocates any tax on business is trying to fool us into believing we are getting a free ride. Should we trust anyone who attempts to deceive us by hiding the taxes we pay?

TAX THE RICH?

It is time to dissect the politicians other favorite – tax the rich. Will it bare careful analysis any better than taxation of businesses? When we throw taxes at the rich, do we throw rocks or boomerangs? The politicians' desires to tax the rich – or at least to appear to tax the rich – are rather obvious. Rich can be defined in many ways. Some politicians claim rich begins with incomes of $75,000 per year or less.

When the rich use money to pay taxes they can't spend it, invest it, or give it to charity. When we tax the rich, we harm the people who benefit from whatever other uses the rich would have made of that money.

No one I know in the $50,000 to $100,000 range acts rich or claims to be rich. They are likely to scoff at the idea that they are rich. Be that as it may, incomes of the majority of people fall below $50,000. Thus, taxing the rich has lots of political charm, even if "rich" begins at $50,000. Getting reelected involves pleasing the majority of voters. Few things please voters more than having someone else pay the taxes.

Taxing the rich involves many a slip 'tween the cup and the lip. The luxury tax imposed on big ticket items in 1990 is a classic example. This tax hit purchasers of expensive boats, and the like, with a 10 percent charge that was to raise billions of dollars. The one thing the politicians overlooked is that rich people – even under the politicians modest definition of rich – don't get their money by being dumb. When the tax went up the rich stayed home. Sales of expensive boats and other taxed items plummeted. The tax revenue never materialized. Not only that, the businesses that made the goods fell on hard times. They now paid less tax instead of more.

The biggest disaster was that those not so rich people who made the boats and other toys for the rich lost their jobs. As a result, they too paid less tax. Unemployment benefits paid to these workers exceeded the revenue raised by the new tax.

Did anyone benefit from this tax? If there were any beneficiaries, they were most likely among the rich. What did they do with the money they didn't spend on boats? Perhaps they invested it and grew even richer. If the tax reduced spending and increased savings, it wasn't all bad. I suggest that you don't argue those benefits too loudly to those who lost their jobs.

One dismal failure doesn't mean we should give up on taxing the rich. Our country's first attempts to launch a satellite failed. This didn't doom the space program. Before damning or praising tax the rich schemes, we should dig deeper.

What do you do with your money? You may pay taxes, spend, save, or donate to charity. You have no other options, unless you have compulsions to burn it or shred it. What do rich people do with their money? The same things. They just do more of it. The distribution among the alternatives may be different.

There is one unalterable principle applicable to all money, no matter who it belongs to. Money, like flash bulbs, can only be used once. The only revocable decision is the one to save. Saved money can later be used for any of the other purposes – but only once.

When the rich use money to pay taxes they can't spend it, invest it, or give it to charity. When we tax the rich, we harm the people who benefit from whatever other uses the rich would have made of that money.

People of modest income spend most of their money. We rely heavily on high income people to invest the money to build the factories, offices and stores that provide a livelihood for us all. The one thing we can be sure of when we tax the rich is that it will have consequences for lots of people who are by no means rich. Some of those consequences may be very unpleasant.

Rich people, however we define them, should pay taxes like everyone else. To avoid falling for political huxterism and shooting ourselves in the foot, or some even more vital organ, we must ask more questions and get more answers before we sign on for some new tax scheme that we are told will hit only someone else.

In the game of taxation it is very difficult to throw fast balls. Mostly we throw curve balls and boomerangs. Unless the voters gain at least some understanding of the complexity of taxation, we will keep on hitting ourselves in the heads.

A Marmaduke World

I have commented on books, opined about TV programs and ventured into the world of sports. Up to now, I have not reviewed a comic strip. Readers, please place your seats in the upright position, buckle your seat belts and hang on. If the review should crash in the ocean, you may use this book as a flotation device.

Well over a decade ago I quit reading the Marmaduke comic. My reason was simple. I didn't find that single panel sufficiently entertaining or enlightening to justify the two seconds it took to read it. The only lower rating would be to say I found it offensive.

Not having to justify my action to anyone, I didn't contemplate why I found this offering so unattractive. I would have found it difficult to justify spending much time pondering my decision. I could have quickly used up all those seconds I saved. They add up to only about ten minutes per year.

It's the supporting cast, not Marmaduke, who are symbolic of what is wrong in America. For so long as millions of Americans wimp out and make accommodation with irresponsible living, the downhill slide of our civilization can only continue.

After a decade or two I revisited Marmaduke for a second look. I discovered Marmaduke was worse than boring. It was down right annoying. Marmaduke personifies (dogifies, if you prefer) irresponsible living. The supporting cast are a bunch of wimps who put up with untold annoyance and inconvenience rather than challenge Marmaduke's arrogant irresponsibility.

When Marmaduke pushes his way into your chair, move over. When he lays in the doorway, bridge over him. When he drags the household furnishings to the doghouse, complain a bit. If Marmaduke were a child, he would be a spoiled brat growing into a juvenile delinquent and worse. Is it Marmaduke's fault? He does what comes naturally (in comic strip land at least), and gets away with it. I find it difficult to laugh at the flaunting of irresponsibility, or at the wimps who condone it.

It's the supporting cast, not Marmaduke, who are symbolic of what is wrong in America. For so long as millions of Americans wimp out and make accommodation with irresponsible living, the downhill slide of our civilization can only continue.

They should send Marmaduke to obedience school. Fortunately the future of our world does not depend upon what happens to Marmaduke. Our future does depend upon what real people do about other real people who are living Marmaduke lives. If we want to live in a Marmaduke world, all we need do is nothing.

SUNBEAMS

Fifty or so years ago when winter melted into spring, I eagerly anticipated the return of the sunbeams. I remember those sunbeams as one of the highlights of my year. Maybe I'm prejudiced toward anything that heralds the coming of spring.

Grandpa and Grandma lived in a traditional farm house from an era already ended. The parlor set at the northwest corner, separated from the dining room by double doors consisting mainly of small panes of glass.

The only way the sun's rays reached the dining room was through the west parlor window and those glass doors. Shortly after the vernal equinox, the sun's northward journey brought it to where, for a few days, the red rays of the setting sun briefly illuminated the dining room.

That short annual appearance was a noteworthy event in itself. The sun alone would have been unlikely to have captivated my young mind. A high backed rocking chair with wide varnished arms set in the dining room with its back toward those glass doors.

How much has the curse of familiarity dimmed and dulled our lives in the age of plenty and instant everything? The worth-while things of life are worth waiting for even if, especially if, they occur but once a year.

At a very young age, and quite by accident, one evening I sat in the rocker as the sun slipped from the sky. As I rocked, a splash of light as red as the setting sun flashed up the far wall of the dining room and back across the ceiling toward me. The harder I rocked the further and faster the sunbeam traveled, painting the usually dim room with cheery brightness. I didn't stop rocking that chair until the sun set beyond the garden taking my sunbeam with it.

Each year as the sun rose in the sky heralding the end of winter, I eagerly anticipated the return of my sunbeam. Whenever I visited my grandparent's farm during this special time of year, I made sure I was in the dinning room waiting when my sunbeam made its brief appearance.

I soon learned enough about the sun and the seasons to believe that my sunbeam should also make an appearance in the fall. It never did. The leaves of the stately maples prevented that encore performance that otherwise would have been a precursor of the autumnal equinox. I was disappointed that my sunbeam performed its dance for only a few brief days each year. I wanted it to come every night.

Had the sunbeam danced every night, How long would I have even noticed? I could not have anticipated its return if it never left. It is unlikely I would even remember it. What I recall fondly as an exciting event of childhood, would have been reduced to a ho hum, soon to be forgotten experience.

How much has the curse of familiarity dimmed and dulled our lives in the age of plenty and instant everything? The worthwhile things of life are worth waiting for even if, especially if, they occur but once a year. We wait for so little that we expect everything instantly and always.

Annual events, like the Fourth of July and the county fair, that an older generation anticipated for the excitement they brought, no longer flash across the heavens of our entertainment drenched world. They are only one of the many sunbeams that we can count on for every evening.

Has our flood of pleasure dulled our ability to enjoy? Has our endless access to thrills and pleasure dulled our sense of wonder and appreciation to where instead of having our excitement limited to a few high points of the year, we have none at all? Today, would even a four year old look forward for an entire year to the return of a sunbeam, or anything else?

WHY DO WE DO IT?

I just finished reading another commentary bemoaning the loss of altruism in our world. How many among us still dream of the days of yore when millions of altruistic souls put the interests of others ahead of their own? What ever happened to that world of self sacrifice? What happened to that wonderful world is that it never existed, never will.

It is beyond the ability of any human to put the interest of any other person ahead of his own. Always has been – probably always will be. How can this be true? Do we not still see selfless acts of compassion and charity? We see acts of compassion and charity. We do not see selfless acts – we never have.

Why then does our world seem more self centered and greedy than in the past? Every decision made by every person is based on furthering that person's self interest. This is a basic, though often denied, principle of life. The reason individual conduct varies so much is, different people are motivated in vastly different ways.

Most of the feelings that guide our decisions are ingrained into us at such an early age that we believe they spring from some law of nature. We think we are selfless because we have been taught and indoctrinated to find the interests of others important to us.

What will you have for breakfast? Your decision will be greatly influenced by what is available. You are likely to also consider how much time you have and the effort required. You may also consider what will be left for breakfast tomorrow, or for other family members today. Considering the interest of other family members, Isn't that a selfless act? No.

When we consider the interests of others, we are in reality considering our own interests. It may be important for us to win favor with another person. Doing things for others may give the doer a good feeling. We may feel guilt if we ignore the interests of others. We may enjoy the success of others. Most of the feelings that guide our decisions are ingrained into us at such an early age that we believe they spring from some law of nature. We think we are selfless because we have been taught and indoctrinated to find the interests of others important to us.

The bottom line is that everything we do is intended to benefit self. The benefits to others may be substantial, still these benefits are only byproducts of the benefactors pursuit of his own self interest. The donor makes a charitable contribution to further the interest of the donor, not for what the gift does for the recipient. The smile on the recipient's face may be the reason for the gift. Still it is the impact of that smile on the giver, not the smile itself, that prompts the gift.

Complaining about self centeredness and trying to change it is as useless as trying to repeal gravity. Individuals behave differently because they have different views of their self interests. The person who expects an immediate and direct benefit from every act appears greedy and selfish. This person is no more selfish than the far sighted individual who sees the importance of other people to his own well being and satisfaction. The later person's decisions are guided by concern for future consequences and the well being of others. This is not because he is less self centered. Rather, because he has a broader view of his self interest.

If we are to change people, we must change their views of their self interests. This is why the early years of a child's life are so important. The views of self interest formed by a child will in large part control his entire life. There is some potential for change. The failures of schools, prisons and welfare programs testify to the difficulties in changing the views of self interest.

Children must be trained – and indoctrinated – at an early age to see actions that benefit others as being in their own interest. Likewise they must incorporate into their lives the importance of forgoing satisfaction of present wants to gain future benefits. Some call this sacrifice. In reality it's investment.

We can't change the self-centered nature of people. We can direct it into constructive channels. Past generations had a far better grasp of the concept that "Me first and now" isn't the best way to pursue our self interests. That's why people of the past appeared more selfless than we. Unless we get back to that broader view of self interest, we have no prospects for a brighter future.

CEREAL WARS

Congressman Charles Schumer of New York called for the Justice Department to launch an antitrust attack on the cereal industry. He concluded that $4.50 for less than a pound and a half of cereal means that the cereal giants, such as Kellogg and General Mills, must be up to something bad. His second assumption, though he didn't specifically mention it, is that the government cavalry must again ride out to rescue the helpless citizens.

One can fairly ask, Why does a pound and a half of cereal, made mostly from grain costing four to eight cents a pound, cost $4.00 to $5.00? Bread made from similar ingredients costs far less. Not only that, bread is delivered fresh to the store every day or so. Any bread that isn't sold quickly ends up selling for a reduced price or being thrown out. Cereal makers don't face the same risks. Keep in mind too, minor brands of cereal often sell alongside the major brands for half the price.

High cereal prices aren't caused by obscene profits. The prices are high because of the incredibly expensive marketing programs.

Does this all mean that the major cereal companies have profits of two or three dollars a box? The short answer is, No. Coupons are a major cause of high cereal prices. The manufacturers, and sometimes the stores, sprinkle those coupons about like rain on an April afternoon. Coupons bring higher prices as surely as April showers bring May flowers. Sellers must increase the prices to cover those discounts, and the costs of distributing and collecting the coupons. Cereal buyers pay those costs.

Some people get reduced prices with the coupons. I challenge any coupon clipper to keep careful records of the time spent collecting, sorting and using coupons, and the money "saved." Keep track of those extra miles driven to the store that gives the best coupon deals. Then figure out how much you are earning per hour. Don't forget that you, and all other customers, would get a substantial part of those "savings" without any effort on your part, simply by eliminating the cost of coupons.

Advertising is the other big non productive cost for cereal companies. Those cereal commercials cost tens of millions of dollars. Anyone who even has to ask who ultimately pays for those adds should skip this article and get a comic book.

High cereal prices aren't caused by obscene profits. The prices are high because of the incredibly expensive marketing programs. The beauty of these marketing programs is, they do very little to increase sales. Mainly they cancel out the marketing campaigns of the other cereal companies.

There is little the Justice Department can do about this situation except institute costly litigation. The cereal companies' costs of the litigation will then be added on to the already high prices of cereal. The government's costs will chew up more of your tax dollars. Except for those who get a tremendous thrill from watching a cereal company get kicked in the shins, the whole exercise just isn't worth it.

Consumers who are upset about the prices of cereal can do what I do. A $4.50 box of cereal kills my appetite. There are brands that sell for far less. Buy them. A 20-ounce box of a kind of raisin brand, which I find at least as desirable as the major brands, sells for under $1.70. The "minor" brands provide fewer varieties. Those willing to pay a couple of dollars a box for a broader selection are welcome to do it. If you choose this route, don't complain about the prices. It's your choice. If sales of low priced brands increase, they will add more varieties. That's how free enterprise works.

Some brands taste different. It only costs a couple of dollars to find out if you like the cereal. That's much cheaper than getting government involved. Asking government to fight the cereal wars is another example of asking government to act where failure is guaranteed, and where consumers have the solution in their hands, if only they will use it.

If sales of major brand cereals fall even 15 or 20 percent, the companies will find ways to cut the prices. Motivation is all it takes to make the marketplace work. Only the consumers can provide that motivation. That's what competition is all about. Ask not that government do what we can do for ourselves.

WHICH WORKS THE BEST?

Michigan's state-wide burning ban just ended. It must have worked. There were far fewer wild fires in Southern Michigan during the 30-day ban between April 15 and May 15 than during the preceding 30 days.

Every spring we run the gauntlet of the fire season. When dead stuff that wintered under the snow bursts free, it drys quickly and lays there waiting for a flame. Once the new green growth permeates the tinder, the fire season is history.

The highest fire risk is when we have a burst of warm dry weather between snow melt and the green season. In these parts that happens, if at all, between the first of March and the end of April, usually in March. During the first two weeks of May it would have been a challenge to start a grass fire with gasoline.

Government has a way of getting out of hand when no one watches it. It's a whole lot easier to look over the shoulder of local government than to look over the shoulder of Godzilla in Washington or a state capital.

It is indeed rare that we aren't well into the green by mid April. Of course the Upper Peninsula, as likely as not, is still fighting off snow. Just like potato planting, the fire season comes later to the North Country. Enter the geniuses in Lansing. They impose a state wide burning ban that begins after the Southern Michigan fire season ends. Is government always stupid?

Even stupidity has its logic. Our bureaucrats, who no doubt live like moles deep inside the Lansing catacombs where they couldn't smell smoke if their own building burned, calculate that most fires in Michigan occur between April 15 and May 15. Doesn't it make perfect sense to ban burning at the time most fires occur? Why bother those little brains with where and why the fires occur?

Another ban comes to Michigan with the changing of the seasons. During the spring thaw we ban heavy trucks from the lightly built county roads. After the roads thaw and stabilize the load restrictions are lifted.

Unlike the burning ban, load limits are established by county government. Restrictions usually first appear in the southern counties and work their way up state as the northern roads soften. The weight restrictions may begin in January and last until April. Sometimes they don't begin until March. Somebody checks the weather and the roads. Each county tailors the load limit period to local conditions. They don't base the load restrictions on last year's weather.

Is it unreasonable to suggest that burning bans established by local government, based on existing conditions, would make more sense and be more effective than the mindless efforts of the bureaucrats in Lansing? Might this be one more example of how the bigger government gets the dumber and less effective it is?

It is fair to ask, Why does local government work better? Load limits are inconvenient and costly to those in the trucking business. If the county Road Commissioners impose restrictions too soon, or allows them to linger after the frost melts, they are going to hear about it big time. The truckers look over government's shoulder.

Government has a way of getting out of hand when no one watches it. It's a whole lot easier to look over the shoulder of local government than to look over the shoulder of Godzilla in Washington or a state capital.

If Godzilla doesn't take kindly to your watching over his shoulder, you are likely to end up between the two halves of the bun for a monster burger. When the big fellow says jump, you better say, How high? Any government big enough to ignore or intimidate the people it's supposed to serve, is too big.

THEY JUST DON'T GET IT

An organization called the DADS Foundation issued a press release regarding Congressional action on funding of anti drug programs. The press release seemed supportive of ending the current federal programs and replacing them with block grants. In case you haven't noticed "block grant" is rapidly becoming the most used phrase in Washington. I'm not sure who coined the phrase. I suspect though that "block grants" are named in honor of the block head who came up with the idea.

A federal grant, is a federal grant, is a federal grant, no matter what you call it. Sure, federal grants are better than programs run by the federal government – about to the same extent it is better to have a leg cut off below the knee, rather than above. The federal government will quit attaching strings to federal grants sometime after fishermen stop attaching lines to their hooks.

There is only one constructive role Washington can play in cleaning up the mess it has worked so hard to create. That is – Get out of the way.

Thus, I give little credit to the DADS Foundation for favoring block grants over federal programs – particularly when most of the press release was devoted to describing the strings that should be attached to the grants. How is giving out a federal grant and telling the recipient how to spend it better than Washington hiring its own bureaucrats to run the program? If anyone figures out the advantage, please let me know.

Money from Washington without any strings doesn't work either. It isn't free money. Still that is the way the recipients usually see it. It's free to them because they didn't earn it. One basic rule of life is, people spend the money they earn far more carefully than the money they don't. What this all means is, the true equation for restoring responsibility to our world doesn't contain a "W."

There is only one constructive role Washington can play in cleaning up the mess it has worked so hard to create. That is – **Get out of the way**. It is understandable that Washington politicians and bureaucrats do not find this to be an acceptable answer. They all have a vested interest in continuing business as usual. Besides, they have convinced themselves that only they know what's good for us. Only they can save us.

The DADS Foundation lists its address as Kalamazoo, Michigan. (Is there any other Kalamazoo?) Kalamazoo is a long way from Washington. I would like to believe that people in Kalamazoo have the confidence to go it on their own without Washington holding their hands and leading them across streets.

The last paragraph of the press release reads:

> "If Congress replaces the current safe and drug free schools act with a carefully designed prevention block grant, the disturbing recent trend of increasing drug use among kids can be reversed. If Congress eliminates drug prevention without any plan, drug use among youth will continue to rise and state and local bureaucracies will continue to grow."

This certainly is a ringing endorsement of Washington's claim that only the feds can save us. Those good people from Kalamazoo clearly accept that they, and all the rest of us, are the bunch of ignorant boobs Washington says we are. If Congress doesn't save us, we are lost. The DADS Foundation is entitled to this belief. All I ask is that when they speak with their voice of failure, they don't paint us all with their brush of incompetence and despair. If the voices of the DADS Foundation truly believe what they say, may I suggest that they disband and go back to sleep. They are the problem, not the solution.

If It's Free, It Must Be Good

I stopped on my way to Kalamazoo and bought a cup of coffee. This may seem to be a rather unremarkable event. Might I point out that I don't drink coffee, or any other members of what I call the "foul brown beverage family." The family includes cola, root beer, Dr. Pepper (I think it's brown), and hot (or cold) chocolate.

Neither do I drink that rather popular beverage that commonly comes in brown bottles. Perhaps that's based on guilt by association. I never set out to avoid beverages because of their brownness. A few years back while mulling over the things I didn't drink, I discovered that I generally avoided all members of the Brown family.

I avoid consuming these popular drinks because when I first tried them I found they had unpleasant tastes. Why are my taste buds so different? I doubt that they are. My recollection is that most others choked these wonderful beverages down until they acquired a taste – and perhaps an addiction. Being of a somewhat independent nature, I refused to play the game.

"Free," the most wondrous word in the English language. Whenever you hear the word free, it should flash like a strobe light and wail like a siren. It is usually a warning that someone is lying to you in an attempt to separate you from more of your money."

I must confess that I was once a social drinker – of coffee. My small streak of compassion prompted this sacrifice. For many people it's a given that everyone drinks coffee. The hostess would not ask if I wanted coffee, rather – How or when do you want your coffee?

"I don't want any," wasn't an acceptable reply. The hostess would likely respond with a blank stare, followed by stuttering and stammering while she tried to think of an alternate beverage. She would also refuse to accept my assurances that I didn't need anything to drink. The hostess might even seem offended.

I use the word "hostess" to maintain historical accuracy. Men were much more gracious and accepting when I declined offers of coffee. If reporting the facts makes me sexist, then paint me sexist.

My desire to abstain from coffee led to awkward and seemingly unending confrontations. Thus, I sometimes accepted the coffee, laced it with sugar and choked a bit of it down. On reflection I doubt that I did this for any reason of compassion. I just grew tired of the confrontations.

Eventually I decided there was no reason why I should accept the hostess's problem as my problem. The hostesses would have to learn to deal with their problems about my distaste for coffee. The last sip of coffee passed my lips many a decade ago.

So why on earth did I stop and buy that coffee? Am I prematurely senile? I'm sure you can find plenty of witnesses to support the senile theory, and use this dissertation as Exhibit A for the prosecution. My reasons for buying that cup of coffee are not so simply explained. I might add, I didn't drink the coffee. I didn't even see it. Now someone is saying, He really is senile.

I bought gas. After I paid the clerk $19.00, she smiled and said, "Would you like a cup of coffee. It's free with the gas." "Free," the most wondrous word in the English language. Whenever you hear the word free, it should flash like a strobe light and wail like a siren. It is usually a warning that someone is lying to you in an attempt to separate you from more of your money.

The first law of economics is eloquently stated by that old adage, "There ain't no free lunch." Nothing is free. Everything is paid for by someone. The most you can hope for is that someone else will pay your bill. I'm not so optimistic as to believe someone else paid for my cup of coffee. The cost of all those cups of coffee was added on to something sold at the gas station. I have a feeling that some of that cost showed up in the price of gasoline. (Note to merchants: I never again stopped at that gas station.)

If my cup of coffee cost $.19, that was a penny per gallon (per gallon of gasoline, not per gallon of coffee). What if my coffee cost $.38? Sure, a penny or two a gallon isn't much. Still, I would far rather keep those pennies and let those who want coffee pay for it.

In the case of my cup of coffee and many other "free" items, the real meaning of the word free is, "You are going to pay for it whether you want it or not."

LESSONS FROM OTHER LANDS

Millions of American citizens languish in poverty and despair. Is America no longer the land of opportunity? Politicians and pundits constantly bombard us with the message that we must create jobs for the unemployed. Why do millions of foreigners risk everything they have, including their lives, to migrate to our hopeless land? How have we so failed to get out the message that opportunity in America is dead?

How can a foreigner in an alien culture – cut off from family and friends, unable to even speak our language – succeed where millions of our citizens wallow in despair? This perhaps is the most important question of our times. Millions of those foreigners do succeed in our land of opportunities lost. A few immigrants end up on welfare. The success stories far exceed the failures.

Job attitudes are far more important than job skills. If individuals are willing to work and learn, there are lots of employers who will teach them. Bad attitudes, not lack of skills, hold back most of the chronic unemployed.

While millions of Americans scorn low paying nasty jobs, the immigrants see these same jobs as opportunities. They take the best job available to an alien who doesn't speak English. They are not taking jobs from citizens. They take jobs scorned by citizens. Most of these immigrants work hard, learn the language and move on to better things. Many build their own businesses. For some success awaits the next generation when the children of the immigrants rise to levels of which their parents never dared dream.

Why don't impoverished Americans follow the same path to success? The journey to success is a climb up a stairway – not a ride on an elevator. No one ever climbed stairs without taking that first step. Until poor Americans are willing to take that first step of opportunity – the nasty low paying job – they will never make it on to the stairway of success.

The welfare elevator stalls before it rises above the sub basement, leaving its riders trapped between floors unable to even see the stairway. Willing, able workers – not politicians – create jobs. The more people work and spend, the more jobs there are.

If impoverished Americans are to again share the American success stories that are now written mainly by immigrants, we must push them off the welfare elevator and point them toward the stairway. Job attitudes are far more important than job skills. If individuals are willing to work and learn, there are lots of employers who will teach them. Bad attitudes, not lack of skills, hold back most of the chronic unemployed.

Increasing minimum wages rips away the first steps of the stairway to success. Without those first steps, entry to the stairway requires a big jump. Those who cannot make that jump are forever denied entry to the stairway of success.

WHEN FREEDOMS COLLIDE

For every freedom there is a corresponding burden. For even one person to be free, all others must bear the burden of that freedom. In order for anyone to have free speech, everyone else must bear the burden of refraining from interfering with that speech. They must also endure whatever you may say. You must bear this burden too, if others are to have free speech. When those who exercise free speech speak irresponsibly, they lend credibility to demands for limits on free speech. These same principles apply to all freedoms.

A young woman made a big ruckus over her alleged right to attend any college she chooses. She demanded the freedom to attend the Citadel, an all male school. If this woman has her way every male in this country will lose the freedom to attend an all male school. In the next logical step, all females in the country will lose the freedom to attend an all female school. Is this young woman's freedom to attend any school she chooses so sacred that 250,000,000 Americans, and their offspring forever after, must be deprived of the freedom to attend a single gender school if they so desire?

When we so desperately need to find social stability, attacks on our few remaining stable institutions are in reality attacks on the foundation of our civilization.

Every indication is that this young woman only wanted to attend the Citadel to make a point. She wanted to prove she could get in. There is also reason to believe she was thinking about the sale of book and movie rights to her story. There used to be a name for people like this, "spoiled brat." Must all institutions and customs that have served our civilization well be destroyed to please a few troublemakers? If the young woman lacked educational opportunities at other schools, her cause would have some merit.

The underlying cause of all the social chaos that flows across our nation is the loss of stability in our world. The environment in which we live changed far faster than we adapted to those changes. Change for the sake of change is worse than foolish. When we so desperately need to find social stability, attacks on our few remaining stable institutions are in reality attacks on the foundation of our civilization.

The young woman seeking to destroy single gender education may have acted out of youthful foolishness. Many of her supporters have no such excuse. They are no longer young, which only serves to illustrate that foolishness is not confined to youth. Those who choose to believe single gender education is an evil that must be stamped out are free to so believe. Why should all the rest stand idly by while one more of our freedoms is trampled to appease the troublemakers and destroyers among us?

Survival of single gender education isn't crucial to our civilization. The attack on the Citadel is only the latest in a seemingly endless succession of terrorist raids on our institutions and customs. Institutions and customs, like top soil, can be eroded away in a single storm. It takes generations to rebuild either one. Civilization cannot survive without customs and institutions, any more than agriculture can survive without top soil. Farmers know that poor soil is better than none. Even if our institutions and customs are not the best possible, we must maintain them until we develop something better. If we destroy what is left, we will be doomed to fall into the anarchy of total individualism. This anarchy will in turn lead back down history's familiar road to total tyranny.

THE MANY FACES OF COMPETITION

It is popular in our day to denigrate and demean competition. Do those who compete against competition understand the creature that they attack?

Competition has many faces. A runner may compete against a fixed goal, such as endeavoring to run a four minute mile. That same runner may compete against himself seeking to exceed his personal record. At a track meet the runner competes against the field, seeking to be first across the finish line. Even while seeking first place, the runner may still be competing to achieve his other goals.

Track and baseball exemplify the type of competition we should encourage off the athletic field. In life we should not impair our competitors' abilities to perform. We should only try to outdo them.

If running a four minute mile is the athlete's all consuming goal, he may win the race at 4:02 and still feel defeated. If the runner is consumed with the desire be first, second place at 3:48 will be a burning disappointment. If the runner comes away seeing himself as a loser, it is his own choice. He is the one who set his goals. Those who demean competition because not everyone can be first, or hit a home run, or get an A, etc., etc.; are simply too small minded to appreciate the unlimited opportunities for personal victory through competition. The person who cannot be motivated and stimulated by competition, and find the satisfaction of victory, simply has no imagination – and no future.

Even when we view competition in the simplistic form of an endeavor to beat an opponent, it still has many faces. In football each team attempts to win by impeding the performance of the other team. Though not quite officially sanctioned by the rule book, part of that attempt to impede the opponent involves physical and emotional impairment of the opponent until he is no longer effective on the field.

Track meets lie at the other end of the spectrum. Each competitor does his best and hopes it is good enough. Any attempt to impair an opponent's ability or interfere with his performance quickly leads to disqualification.

Baseball falls between football and track. Each team tries to frustrate the goal of the other. Containment of the opponent is accomplished by exceeding the opponent's effort, not by trying to impair that effort. The pitcher tries to get the batter out by throwing an unhitable pitch, not by having the catcher grab the bat. Tripping base runners is also a no, no.

By trying to cripple others to achieve our own victories, we harm the others far more than we benefit ourselves.

Track and baseball exemplify the type of competition we should encourage off the athletic field. In life we should not impair our competitors' abilities to perform. We should only try to outdo them. By outdoing others we don't harm them. By trying to cripple others to achieve our own victories, we harm the others far more than we benefit ourselves. Is the runner who wins by tripping his opponent a better runner than if he finished second without the tripping?

Perhaps our taste in sports is a barometer of the decline of America. The honest competition of track and baseball now takes a back seat to the destructive competition of football and similar sports. Hockey can be played either way. Still, the hockey games that draw the most fans and generates the most enthusiasm, are those where players brutally impair their opponents by means both fair and foul.

If Americans' taste for crippling opponents ended in the sports arena, it wouldn't be of great concern. Unfortunately what we see at the stadium is a manifestation of the character – or lack thereof – of our nation. The same attitude extends to all endeavors. Until you see Americans' taste in sports turning away from football and violent hockey, and back to the likes of baseball and track, we aren't going to see a kinder, gentler America.

COMPETITION – SAINT OR SINNER?

For more than a generation we have heard bad things about competition. Particularly, we hear that children shouldn't compete with each other. The competition bashers want to do away with grades in school. They don't want anyone to try to out do anybody else.

Is competition bad? Certainly there is a negative side to competition. Unrestrained competition can be destructive. Food poisoning kills. Bad water can be just as deadly. Contaminated air kills too. We don't try to end eating, drinking and breathing because of the hazards. We recognize that bad air is better than no air at all. We try to clean up bad air rather than eliminate it. Should we try to eliminate competition because it's contaminated?

Competition is as essential to the survival of our civilization as air and water are to life. The only difference is, death comes more slowly from loss of competition.

People who do nothing, contribute nothing to the survival and prosperity of the human race. People do nothing, unless they are motivated to do something.

What motivates people? Fear is the greatest motivator. Fear of impending doom has been the main force driving the human race. For most of history most people lived barely a step from extinction. They competed with each other for the necessities of life. Competition is as old as breathing.

When the successful efforts of a civilization allowed it to draw back from the edge of destruction, the people didn't immediately lay back and rest. The competitive drive ingrained into them drove them to continue their efforts although the threats to survival were now distant. The competitive energies propelled the civilization to new heights. In time though, members of the society began to question expending all that effort, which in a seemingly secure environment appeared unnecessary.

Competition and productivity were replaced by relaxation and pleasure. While people lived off the fat of the land, the spirit of competition eroded until productivity fell below the level needed for survival. Like an overweight person on a diet, the society lived off its stored energy. After expending the stored energy, the civilization fell into chaos and again battled for survival. Many didn't survive the transition back to the harsh world of competing for survival.

If we are successful in destroying the spirit of competition, our civilization will go the way of all those before us. Without the spirit of competition, we will slide to the edge of the abyss of destruction. Finally motivated by the threat of extinction, it will be too late for our way of life and most of the people. Those who survive will reinvent competition.

Competition is as essential to the survival of our civilization as air and water are to life. The only difference is, death comes more slowly from loss of competition. Those who try to eliminate competition are trying to put a gun to the head of our civilization and pull the trigger. We should fear these people more than any serial killer who at most kills a few hundred people. Those who seek to destroy the spirit of competition, seek to kill us all. Maybe they mean well. That will be small consolation if we allow them to succeed.

CONFUSION AIN'T ALL BAD

For the past sixty years America marched toward an elusive goal. Though the march wavered at times, there was always a consensus as to the general direction of the march. Government had a major role to play in the lives of all of us. It was the primary role of government to protect and care for the people. Government must care for those who can't or don't care for themselves. Mainly the federal government was to be the care giver.

A minority always questioned the direction of this march. These hardy souls were branded as cruel, heartless and uncaring. How could any decent human being be against helping the helpless? The definition of helpless quickly expanded to include most everyone. To question that government must help was more socially unacceptable than passing gas in church. The only question acceptable in polite society was – How can government help people? Those faint voices that whispered, "Government isn't helping," were shouted down as only trying to justify their own greed and hardheartedness.

History tells us, Washington demands that we all blindly march in lock step down the same uncharted path. When we all tread the same path, we all risk falling off the same cliff.

The leaders of the great march have lost their compass. It is no longer heresy to question the direction of the march. Frightened advocates of big government still attempt to brand as heartless and mean spirited those who question government. These brands no longer burn the way they used to. The marchers now mill around and wander in confusion. Though millions now see that we were marching in the wrong direction, few truly understand what went wrong or why. Even fewer have any idea how to find a new direction for our march.

Everywhere we hear shouts of go this way or go that way. More police and prisons, less welfare, more midnight basketball, more money for education, capital punishment, crack down on pornography, etc., etc. There is no consensus. There is also little if any hard evidence that these new directions will fare any better than the old one.

It is good that we have fallen into confusion. When lost in a swamp, marching with certainty in the wrong direction isn't better than not marching at all. The confusion we see about us today is progress. We now know we are lost and have started to seek new direction. When the march resumes, we have no guarantee it will head in the right direction. At least we have a chance.

Now is the time to ponder, not to shout. We have no guarantee that the loudest voice will proclaim the way of salvation. It is even less likely that the easiest direction, or the one that feels best, will be right either. Before we march in any direction, we must send out scouts to map the path. This is where the federal government is all but certain to fail. History tells us, Washington demands that we all blindly march in lock step down the same uncharted path. When we all tread the same path, we all risk falling off the same cliff. We are also denied the opportunity to see where other paths might lead. These are compelling reasons for disbanding the federal bureaucracy. With all the uncertainty about the direction we must march to restore our civilization, we can ill afford to all march in the same direction – no matter how right it may feel. We must follow different voices until someone finds the real path out of our swamp.

We must attack the federal bureaucracies with the spirit of wild eyed, fire breathing radicals. The only certainty we have is that once we decimate the federal bureaucracy, whatever springs up at the private, local and state levels will be better than what we tore down. The new growth being smaller and less deeply rooted, will also be much easier to guide and direct in productive directions.

For now we should spend little time concerning ourselves with what will replace the federal bureaucracies. We must concentrate on ridding ourselves of them. When your stomach is full of poison, it isn't time to worry about where you will get your next meal, if you get rid of the poison. Poison isn't better than no food at all. For starters we should abolish, not reform, entire departments such as Education, and Housing and Urban Development. It isn't that these departments don't deal with important matters. The problem is, the federal government has no business being involved in those matters. All it does is march us deeper into the swamp.

WAR!

The 1950s witnessed the sowing of the seeds for a long and bitter war. In the '60s those seeds germinated into open combat. Certain of victory, we forged ahead. Why should we fear a disorganized enemy with resources and manpower a small fraction of our own? We would squeeze our enemy. We would deprive him of supplies. Without the continuous flow of those supplies the enemy would cease to be a threat. The war would end with a whimper.

It was not to be. Our enemy responded with cunning and resourcefulness to every escalation we made against his supply routes. We trumped each of our victories. We gloated when we seized or destroyed a mass of those vital supplies. We even destroyed entire units of the enemy's forces. Replacements flowed in from the shadows. Our efforts seemed more to strengthen our enemy than to weaken it.

With time, we began taking note of civilian casualties. Combat spilled across national boundaries threatening the stability of neighboring nations. We went so far as to invade one of those nations to further our war efforts.

Early on, a few voices spoke out against this no win war. They were ridiculed and shouted down in our push for victory. Our irrational fear of the consequences of defeat drove reason from our minds. Our civilization would fall if we stopped the war. Millions who understood little of the war, or how it was to be fought and won, never wavered in their certainty that we had no alternative. So, even now in the mid '90s, the war to cut off the supply of illegal drugs rages on, even further from victory than three decades ago.

We seem unable or unwilling to see the real cost of this war. If we were only failing to win – we could live with that. The death, devastation and destruction from this war far exceed those caused by the drugs we fight. Spending hundreds of billions of dollars has done nothing to cut off the supply of drugs on the streets. When one drug lord falls, a dozen pretenders stand ready to take his place. Even if we could seal our borders to cocaine and opium; even if we could destroy every coca leaf and opium poppy in the world – we would have accomplished nothing. Domestically produced substitutes already wait in the wings, ready to fill the void.

The billions of dollars from the illegal drug traffic corrupt police, other public officials, and businesses. Those same dollars threaten to destroy or dominate the legitimate governments of Central America, South America and the Caribbean. Raging wars and gun battles destroy our own neighborhoods and overflow into our schools. This war preoccupies our police, clogs our courts, and fills our prisons.

In response, the blind and opportunistic among us band together to federalize the criminal law. They care not that there isn't one shred of evidence that federal law enforcement is more effective than state and local enforcement. Federal laws require federal police to enforce them. We pride ourselves on not having a national police force. We see how national police forces in other countries become vehicles for tyranny and oppression. Yet, in our blind pursuit of the war on drugs, we are well on the way to creating that national police force that can be used by any would be despot to destroy what is left of our individual freedom.

Even if we could seal our borders to cocaine and opium; even if we could destroy every coca leaf and opium poppy in the world – we would have accomplished nothing.

Are drugs really so frightening a menace that we must continue this war unabated until it destroys us? The Association of the Bar of the City of New York conducted a study of the war on drugs. The committee faced up to the total failure of that war and recommended the repeal of all federal drug laws. One finding was that only 2 percent of Americans say that drug laws are stopping them from using drugs. In other words there is no reason to fear that ending our foolish efforts will create an avalanche of new users. The report of the New York Bar can hardly be written off as self serving. How many tens of millions of dollars are lawyers on both sides of the battle front making from continuing the war? Lawyers, bureaucrats and drug lords are the only winners in this war. We are all losers.

The only effective drug war we have fought is the one against tobacco. Tobacco use is still high; however, it has decreased substantially from the level of three decades ago. Was this success from a major law enforcement program? No way. The success grew from a long steady flow of information and education. If the war on drugs is ever to succeed, it to must be fought from the soap box and the pulpit – not from the police station and the courthouse. As in all social battles, victory requires winning the minds and hearts of the people. Seizing their bodies will not work.

Remember how our government lied to us and deceived us about the Vietnam War? Should it come as a surprise that government uses the same tactics to justify the anti drug empires and billions of dollars being pumped into the war on drugs? We hear what a great percentage of high school seniors have used drugs. We hear how many millions of people use drugs.

We are conditioned to think of drug users as mindless hulks with vacant stares, and arms pitted with needle marks. This is the equivalent of seeing a skid row derelict as the typical user of alcohol. What is rarely mentioned is that most of those counted among the users, tried drugs and quit. Even most of the active users are casual, occasional users, not hard core addicts. Drugs are not as fiendishly addictive as we are indoctrinated to believe.

The drug related death of an athlete or movie star reverberates through the news for years. Drug deaths make such big splashes in the news because they are so rare.

The number of people dying from drug use is infinitesimal compared to the numbers killed by either tobacco or alcohol. This doesn't make drugs good. If we could banish drugs by force, without destroying ourselves in the process, I would be all for it. I am also all for facing reality. Few crimes are the result of drug users going berserk. Most drug related crimes spring from raising money to support a habit, or from battles between drug dealers – and from the bribes and general corruption used to maintain the drug empire. Lots of innocent bystanders suffer and die. Besides this, we risk the subversion of our government and creation of a police state.

Is it really worth all this devastation to fight the good fight to <u>try</u> to stop a small percentage of our citizens from frying their brains and bodies? Remember, trying is all we are doing. What is the name of that famous road paved with good intentions? On the bottom line in the war to cut off the supply of drugs, there are no victories. Those who want drugs still get them, even if they have to kill to do it.

Drug use does not threaten to destroy our civilization and way of life – the war on drugs does.

Alcohol comes in many forms but is still a single substance. Drugs are far more complex. Blanket repeal of all drug laws isn't the answer. We must find a way to wean drug users from the illegal supply, thus pulling the rug from under the criminal element. God forbid that we follow the government's approach to gambling, and start advertising legal drugs.

Limited legalization will not end drug use any more than the repeal of prohibition ended drinking. Repeal did cut the legs out from under the bootleggers and their illicit empires. It also took the machine guns off our streets until the drug lords brought them back.

We don't have to legalize everything for everybody. Drug sales to minors should remain illegal. We should let the states handle the details. This way we will not subject the entire nation to the same experiment. Yes, it would be an experiment. When seeking a solution to a problem that up to now has proved insoluble, experimentation is the only possible approach. One of the big problems with federal programs is that they usually subject the entire nation to the same experiment, and try no others. The alternative is to continue to pursue proven failure. Is it rational to pursue guaranteed failure just because the certainty of success of the alternatives has not been proven? They never will be proven if we don't try them.

Limited legalization will not end drug use any more than the repeal of prohibition ended drinking. Repeal did cut the legs out from under the bootleggers and their illicit empires. It also took the machine guns off our streets until the drug lords brought them back. How grievous must the self inflicted wounds become before we stop the madness? Treatment, education and enlightenment may or may not cut drug use. Only time can tell. How can we do any worse than we have with our futile attempt to cut off supply? The social cost will be infinitely less, even if we can't put a dent in drug use. It's time we tried a new approach. In our attempt to cut drug use it is far better to follow the successful example of the campaign against smoking, rather than the failed battle plan of the Vietnam War.

IN PRAISE OF PAIN

Few endeavors preoccupy our world more than the avoidance of pain. How many millions of dollars worth of pain killers pass over the counters of our nation's drugstores each year? The term "pain killer" speaks volumes. Pain is evil. We unashamedly seek to kill it. Oh, the joy of a world without pain.

What if there were no pain in our world? Would we then dwell in paradise? Would it be good to be able to put your hand in boiling water and feel no pain? Would eliminating the discomfort from overeating or overdrinking be a good thing?

> **When we pay property taxes we still feel the pain. The hidden taxes we pay through businesses, and the withholding taxes, dwarf the property tax. We curse the property tax because of the pain it causes, not because of its size.**

I read of a man who due to some defect in his nervous system felt no pain. He nearly burned his hand off without realizing it. Pain may be unpleasant. Unpleasantness sometimes serves a useful and valuable purpose. Pain is a warning that something bad is happening to the body. Pain lets you know that cells in your body are dying. It is then up to you to do whatever you can to stop the damage. To curse pain is to damn the messenger. Pain does not damage your body, it only alerts you to the damage.

Pain, like a baby, doesn't stop screaming just because you got the message. Those screams continue until the problem is solved. It is annoying when the messenger keeps on yelling after you get the message – particularly if you are powerless to correct the problem. That is why we spend so many millions trying to make pain shut up. Also, some people like to do the things that cause pain. They don't like pain to annoy them with the constant reminder that they did something stupid – like overeating or overdrinking.

Not all pain is physical. We are equally eager to avoid mental and emotional pain. Those who inflict mental or emotional pain are not numbered among our favorite people. Politicians figured that out.

Politicians also figured out that taxes cause pain. In keeping with the trend of our times politicians do everything in their power to make taxes painless.

The biggest pain killer introduced into the world of taxation during the twentieth century is the withholding tax. Taking away money we never held in our hands is far less painful than requiring us to reach in to our savings, or to get a loan to pay thousands of dollars at once. This is why people tolerate income taxes and sales taxes while damning property taxes. Property tax is not our biggest tax, it is only our most painful one. When we pay property taxes we still feel the pain. The hidden taxes we pay through businesses, and the withholding taxes, dwarf the property tax. We curse the property tax because of the pain it causes, not because of its size.

Politicians who inflict painless taxes are not our friends. They are no-better than a physician who drugs an athlete before a game so the athlete can destroy his body without feeling the pain. In the end the athlete will be far worse off for not having felt the pain that could have curbed his reckless conduct. The same is true for taxpayers.

We should feel the excruciating jab of every tax we pay. We should suffer and scream each time a tax dollar is torn from us. Only when we again feel the pain of the taxes we pay will we act to protect the body politic from the devastation of the grievous wounds inflicted by excessive taxation.

DEATH AND TAXES

The IRS must die. The foregoing declaration didn't come to me late in the evening of April 15 while battling those cursed IRS forms. No, it was the lead for an article in *National Review.*

Does the evil empire of the IRS have the immortality denied to the communists? Not unless we grant that immortality. There are far less intrusive, and far more efficient, ways to raise the revenue to support the federal government. . .

Reasons for abolition of the income tax and IRS abound. Calculations show that small businesses spend $3.00 to comply with the Internal Revenue Code, for every $1.00 paid in taxes. Hardly a model of efficiency. Estimates show that taxpayers spend $200 billion complying with the Internal Revenue Code that generated $700 billion in taxes. This doesn't include the billions the IRS spends harassing taxpayers.

The 9,400 pages of tax law, and the 4,000 pages of forms are understood by no one. Every time Congress enacts tax "simplification" the Internal Revenue Code grows longer and more unintelligible. The income tax puts the federal government's nose in everybody's business.

The United States has one of the lowest savings to spending ratios in the industrialized world. If it weren't for foreign investments in our country, our economy would collapse. Relying on foreigners to provide the factories and offices in which we earn our livings is both foolish and dangerous. Our income tax laws punish saving and promote spending. This alone is reason enough for radical change. If we are to have a sound economy, we must save more.

Perhaps the greatest indignity inflicted on our nation by the income tax, is perversion of our businesses. Business investments are made based on artificial requirements of tax law, rather than for sound business reasons. Businesses regularly make decisions that would be totally dumb, except for the tax penalties and subsidies which flow from the decisions. The first question any savvy business person asks before investing is, "What are the tax consequences?" All these wrong minded decisions stifle the productivity of our nation reducing the incomes of us all. Elimination of the business income tax will reduce the prices of American made goods, both at home and abroad. This will make American products more competitive. The result will be economic growth and reduction of the trade deficit.

I expect I've said more than enough to convince most people the income tax is a bad thing. Most people don't need much selling on that point. Does talking about income tax do any more good than talking about the weather? Fifteen years ago few dared even dream of the demise of the Communist empire. Ten years ago talk of the fall of communism was still seen as wishful thinking. Then the Berlin Wall collapsed and before our heads stopped spinning communism died.

Does the evil empire of the IRS have the immortality denied to the communists? Not unless we grant that immortality. There are far less intrusive, and far more efficient, ways to raise the revenue to support the federal government, even if we foolishly continue to support it at the present level. A national sales tax would be both less intrusive and less costly. Most states already collect sales taxes with far less expense and intrusion than results from the income tax. These same state agencies could collect the national sales tax and remit it to Washington. We must totally dismantle IRS and the income tax machinery. We can ill afford to run the risk of ending up with both taxes.

The downside would be 110,000 unemployed at IRS and thousands of formerly high paid tax lawyers and accountants looking for new careers. Is that really a downside?

Politicians in Washington are starting to seriously talk about ending the income tax. It isn't going to happen today or tomorrow. It can happen, and sooner than you may think. A little public support for the end of the income tax is all it will take. When we step into the twenty-first century, both communism and the IRS can be road kill in our rear view mirror. This rosy picture will become reality only if we, the people, make our voices heard.

DEMONS

In our age of enlightenment people who talk of demons often draw funny looks, if they draw any looks at all. Passers by may suddenly give in to the urge to walk on the other side of the street. Still the risk of being seen as strange is no excuse for making a left turn away from reality.

A few weeks back when I attempted to start my S-10 it responded not at all. The battery wasn't just weak – it was stone cold dead. How dead was it? It was so dead that the digital clock was comatose. A quick jump from my other truck and all was well. A week or so later the S-10 again mysteriously died in the night.

A few days later, I noticed the brake lights were on after I emerged from the vehicle. There was nothing wrong with the switch at the brake pedal. Even removing all the fuses didn't dim the glow of those cursed red lights. Removing the battery cable did dampen the spirits of the tail gate party.

The most devilish of the spirits toyed with my son's Toyota. (Maybe next time he'll get a real ota instead of a toy one.)

A few days later I undertook to dissect those evil devices concealed inside the steering column. Before I accomplished anything even vaguely electrical the laughing tail lights ceased their smiles. Though I had no idea what wrought the miracle, all is well that ends well.

When my wife arrived home a few days later, she inquired as to why the tail lights were on. Total dissection of the steering column and its magical contents had no impact. I did discover that the four way flashers didn't work, and that I couldn't signal for a right turn. Curses, I'll never be able to sell that truck to Rush Limbaugh.

The beast did respond when I attempted to signal right with the ignition off. The instrument panel lit up like a Christmas tree and the gauges came to life. While signaling right with the engine running, turning off the ignition had no effect. The engine ran on.

DEMONS

The weather was lousy, so I put the S-10 on life support to sustain it until a better day. All night long the battery charger hummed and those tail lights glowed. Two days later when I awoke the cheery glow from the driveway had vanished. The S-10 now behaved as though nothing had happened.

So, you don't believe in demons? Well, if you have a better explanation of what invisible force enters and flees that truck in the dark of night, Out with it. Your wisdom might spare me all those funny looks I'll get when I call an exorcist next time that truck is possessed. In fact, isn't the second possession a repossession – that's something that should never happen to any vehicle, especially one that is paid for.

The S-10 is not the first vehicle in my family to experience demonic possession. My wife's car will not start below 10°. You may say, Nothing unusual about that. The car refuses to respond in any way to a mere turn of the key. Connect a wire between the battery and the right screw on the solenoid, and the car springs to life.

I do not perform exorcisms when the temperature is below 10°, and the demon flees at higher temperatures. Thus, this particular evil spirit is likely to evade identification forever, unless she changes her habits.

The most devilish of the spirits toyed with my son's Toyota. (Maybe next time he'll get a real ota instead of a toy one.) One evening he parked and locked the car. In the morning it set against a barrier on the other side of the parking lot, with scars in the gravel where the rear wheels spun to a stop. The mischievous spirit started the car and drove until it could go no further. Imagine what that demon could do if he reached the accelerator.

Skeptics might question this last account. If you saw all the other tricks that car could do, you wouldn't even be surprised that it went for a spin by itself. It had the original wiring system from Hell.

The Toyota was the first possessed vehicle in the family. Is it possible the Toyota brought the troublesome spirits with it from the Land of the Rising Sun, and infected the other vehicles? That's one heck of an argument for, "Buy American." Maybe Japanese cars aren't really better. Could it just be that no one dares say that they aren't?

WHAT WE ALL NEED TO KNOW

I didn't learn everything I needed to know in kindergarten. In fact, I didn't learn anything I needed to know in kindergarten. The next eight grades were not much more productive. Except for a few oases, high school was a continuation of the wasteland. I didn't fair much better in college. I must have been just plain dumb.

I started to catch a few glimmers of understanding in law school. Still, the world didn't come into focus. So where did I learn? Did I ever learn? Live and learn. Knowledge and understanding are there to be found. Read. Listen. Observe people. To understand the world, to understand all that is worth while, requires understanding people. Until crossing that river the promised land is but a dream.

Rare seeds of wisdom were sown through out my formative years. Unfortunately they were mixed with generous helpings of weeds.

Unfortunately my formal education did little to even lead me to the river. I had a few good teachers that stimulated thought. They taught me to ask, to question, to think. I do not mean to imply that all the other teachers were bad. They did their jobs as they saw them. Mostly they fed me a diet of sterile facts. I was expected to ingest these facts and regurgitate them on command. Fortunately I had a weak stomach for this. Often I only pretended to swallow.

Rare seeds of wisdom were sown through out my formative years. Unfortunately they were mixed with generous helpings of weeds. Often, my mentors showed more interest in nurturing the weeds. When I broke free from the bonds of formal education, I cultivated my own garden. By trial and error I learned worthwhile knowledge could be separated from the weeds. The long slow process still continues. It's difficult to recognize a weed until the seed is ripe.

What is education? The better question is, What should education be? Education is preparation for what's ahead. It is preparation for life. Throughout most of human history the main providers of education were parents and other family members.

Schools were invented when the job of educating became too big for the family. Schools supplemented the family education. They were not intended to replace it. Education involves far more than the accumulation of facts. Facts are useless if we know not what to do with them. We must learn to think. We must know how to solve problems.

Most problems involve people. To solve these problems, we must understand people. We must understand what motivates people. We must understand what irritates people. We must understand how people get along with each other, and use each other. This understanding of those with whom we share the world isn't found in readin', 'ritin' and 'rithmitic. It is sociology, psychology, economics and the like. For many these are forbidding and scary topics. If we received the education we should, these subjects would be familiar, not frightening. These subjects were missing in my education until I reached college. My knowledge in these areas was so woefully inadequate that when the strange new subjects were brought up in college, I was unable to even see their importance. Eventually I started to catch on.

A little understanding of human nature goes a long way toward understanding the problems facing us today. We must understand the problems before we can find solutions. Isn't all this understanding beyond the grasp of most of us? Shouldn't we just leave it all up to our leaders in government? That's pretty much what we've been doing. The politicians and bureaucrats like it that way. For them it's job security. Look at the results.

Government of the people, by the people, and for the people. The politicians work for us – at least they should. We hire and fire them on election day. If we don't understand the politicians' jobs, How can we know if they are doing a good job or bad? If we can't tell what kind of job they are doing, we are likely to end up voting for the best story teller. Do we want stories or results?

Illegal drugs, poverty, crime, education, discrimination, violence and all the rest of the front page problems involve people. The only way to make things better is to change people. For years government has been trying to force people to change. The results have mostly been failures. Even government doesn't have the power to force the changes it seeks.

Force has it's place. One hundred people may force one person to change. If the hundred are up against thirty, don't expect the same result. If the hundred are up against another hundred, forget it.

We must find ways to get people to want to change. Sure, it will be slow. It won't be easy either. It's the only way. As a nation we have only one problem. Many among us do not lead responsible, productive lives. Is it their fault? Whose fault is it? How do we get them to change? These are the questions we must ponder and answer if we are to reverse the deterioration of our society.

The battles in the war for the hearts and minds of people rage about us. There are some successes. There are lots of failures. The politicians and the broadcast media are mostly hawking snake oil. These "miracle cures" for our social ills fail because they ignore the real world and the real people who live in it.

End Welfare in Eighteen Years?

Welfare is one of the hottest political topics of the day. Even the advocates of the welfare state can't claim, with a straight face, that the system works. All the politicians are for "reforming welfare" or "ending welfare as we know it," at least, so they say. Before taking the politicians at face value, pause to remember that politicians love to say what voters love to hear. Also, consider which one of the politician's faces is showing as he speaks.

We must go back to the pre 1960 approach of, either support your baby or give it up for adoption. We cannot afford to recognize any right for individuals to have and raise children they cannot support.

So, how do we end welfare? It is easy to start feeding the birds in the winter. Stop that feeding in January and the resulting disaster will be far worse than anything that would have happened if we had never started the feeding. Any attempt we make to get large numbers of people with children off welfare is all but certain to fail. Converting welfare into make work government jobs and providing taxpayer funded child care isn't a solution. It's just expensive cosmetics for the welfare corpse.

I doubt that there is anyway to get most mothers off welfare without rounding up the children and putting most of them in "orphanages." There is a function to be served by "orphanages." Using them as corrals for the fruits of a mass roundup isn't one of them. Most single mothers have gotten themselves into a mess that they can't escape, even with effort and help. The only way most of these women will ever be able to provide for their children without charity, is to find a wage earning husband. For most of these women, that card isn't in their deck.

So, must we give up? No. The children now on welfare are not the real problem. Nine years from now, half of them will no longer be children. In eighteen years, none of them will be children. If we end additions to the welfare roles, the welfare problem will solve itself.

Denying welfare to new applicants will not require finding ways to provide for masses of partially grown children. There will be a steady – and hopefully diminishing – stream of new borns without parents capable of supporting them. As a nation we can handle that stream though we cannot deal with a flood.

We must go back to the pre 1960 approach of, either support your baby or give it up for adoption. We cannot afford to recognize any right for individuals to have and raise children they cannot support. Particularly, teenage single mothers must not be encouraged or subsidized to raise children. Studies show that children born to teenage girls living in poverty have almost no chance for a decent life, unless they are removed from that environment. The one thing proved to give these children a chance is adoption at birth. Parents who fall upon hard times after the children are born are a different problem that requires a different solution. One size fits all solutions don't fit anyone.

We must make adoption far easier, and final. Attempts to revoke an adoption must not be allowed, except in clear cases of fraud and deceit by the adopting parents. A male not married to the mother should not have any claim to the baby or any voice in the adoption. In addition to simplifying the adoption process, this will send a clear message that society does not condone illegitimate births. If males want the rights of fathers, they have to act like fathers.

The attitudes of society being what they are, it is unrealistic to expect that we can totally dry up the flow of new welfare recipients. If we can cut the flow in half, we are well on our way to a solution. If eighteen years from now we have cut the number of children on welfare in half, that will be a major success. Considering where we are headed now, just keeping the numbers from increasing for eighteen years would be a significant accomplishment. We must not demand ideal solutions at the expense of achievable improvements. Even for those who don't believe that half a loaf is better than none, there is nothing wrong with going after the whole loaf one slice at a time. In the bakery of life this is usually the only way to get the whole loaf.

FREE ENTERPRISE

Free enterprise or capitalism, whatever you call it, is the greatest economic system devised by man. It rewards effort and ingenuity. It encourages production. The free economy produces a standard of living that no other system can even approach. The efficient production of goods and services benefits not only the successful entrepreneur, but also his employees, customers, and all of society. If the free enterprise economy were destroyed, it would have to be reestablished, or else the economic collapse would doom civilization as we know it.

The free enterprise system rewards the ruthless and unscrupulous. The rich use it to get richer by making the poor even poorer. It brings out the worst in people by encouraging and rewarding their greed. Entrepreneurs grow wealthy by using other people.

How many employees don't have the same goal? That is, to get as much pay as possible for as little work as possible, without getting fired?

Which view of free enterprise is right? Is free enterprise a blessing or a curse? Before squaring off with the ultimate question, it might be well to take a brief look at this creature which is so praised and so cursed.

The heart of free enterprise is businesses. Businesses may be large or they may be small. General Motors and Exxon are businesses. So is the shoe repair shop owned and operated by the son of the man who started it from nothing fifty years ago. What is a business? Anytime one or more people engage in exchanging goods or services for something of value, for the purpose of earning a profit, it's a business. The primary purpose of a business is to earn profits. Whether the participants contribute their time or their money to the enterprise, they expect to get something in return. If they expect nothing in return, it is a hobby or just plain exercise. It isn't a business. To curse a business because it seeks a profit is as senseless as to curse a tree for growing leaves or needles. A tree that fails to grow leaves or needles dies. So does a business that fails to earn a profit.

How does a business earn profits? First the business must have income. If income exceeds expenses, what's left over is profit. Our shoe repairer uses his profits to feed and cloth his family. He may reinvest some of the profit in his business. Why would he do that? Only because he believes the investment will lead to more profits in the future. The shareholders who own General Motors do the same thing, only on a bigger scale. They don't make the decisions and perform the work themselves. They hire directors, officers and other employees to do it.

To make a profit a business must have money coming in. How does a business get money? There are only two ways to get money. Someone gives it to you, or else you take it. If you simply take it, you are either a crook or government. When crooks take money, it is called larceny or fraud or something like that. When government is the taker, it is called taxes or fees or something like that. Businesses and charities get money by convincing others to give it to them. Most people expect something in return when they give out money. The return on gifts to charities is usually intangible. The donor may receive satisfaction, peace of mind, public recognition, or appeasement of conscience. Businesses usually have to throw in something a little more tangible to separate people from their money.

Never forget that the purpose of each and every business is to separate people from their money. The less a business gives in exchange for the money it gets, the greater the profits. The ultimate success might seem to be to take in money and give absolutely nothing in return. The big problem with this extreme is that it usually cuts down on repeat business. Most businesses need repeat customers to survive. Thus, the real key to success in business is to give the customers as little as possible and charge them as much as possible, while keeping them happy enough to keep bringing money. Lest anyone start feeling righteous indignation toward businesses, keep the following in mind. The fact that maximizing income is the basic principle on which all businesses operate, does nothing to distinguish business people from anyone else. How many employees don't have the same goal? That is, to get as much pay as possible for as little work as possible, without getting fired?

Free Enterprise

What limits how far a business will go to separate people from their money? There are legal limits. Most business people try to avoid practices that could land them in jail. Generally the need for satisfied customers is a far greater limiting factor. It is not illegal to sell a low quality box of cereal that tastes absolutely awful. The profits per box may be great. Don't expect much repeat business. The store could offer a great tasting cereal for $20 a box. Don't expect much volume on that one either. The main restraint on business practices is competition. That is the strength of the free enterprise system. Self regulation, not government regulation, maintains the quality of products and keeps the prices down. Competition is indispensable to promoting self regulation.

Some people are more unscrupulous and ruthless than others. In no-holds-bared competition, the unscrupulous and ruthless are likely to be the survivors. So, in the free enterprise system the better you are at ripping people off, the more likely you are to get more chances to do it. Yet, a system that does not reward successes and let failures fall by the way side, is doomed to eventually destruction by its own inefficiencies.

Like fire, the free enterprise system is neither good nor bad. It has the potential for both good and evil. Which potential is realized depends on how it is used and controlled. Like fire, if left unchecked the free enterprise system will reek great devastation, and in the end perish for lack of fuel. If properly guided, it can continue to benefit humankind as it has for so long. Destroying the free enterprise system because of it faults and excesses would be as detrimental to all of us as banishing fire from the world because of its potential for destruction and death.

The role of government in free enterprise should be the same as the role of the referee in a wrestling match. Government should act to penalize eye gouging and chock holds. Government should not quash competition. It should prevent no-holds-bared competition by baring the bad holds.

Old Growth

When the European settlers arrived in America, old growth forests spread over much of the continent. The mature trees provided shelter for some animals and slowed the runoff of rain preventing floods. Those trees beyond their prime produced little or no new wood. Mainly, they aged and rotted while waiting to die. Meanwhile, shade from the old trees blocked the sun light, rendering the forest floor almost void of life. Few young trees grew among the giants to replace the elders when they fell.

In its youth free enterprise is an ever changing array of many species of small businesses. The economy brims with the vigor of life. All who share the domain benefits from the youthful vigor.

After fire ravages the old forest, abundant new life and growth fill the void. Wildlife prospers. Young trees produce new wood. The vigor of youth replaces the decline of old age.

The landscape of the aspiring forest changes rapidly. First, weeds and grass provide shelter for animals, and for the seedlings that will grow to soon dominate the landscape. As time passes, the trees overshadow the lesser species reducing them to spindly growth on the forest floor. Most animals disappear for lack of food. Even among the trees, only the strongest survive. In time only a few of the trees of the young forest remain as aging giants in an old growth forest. Then, even seedlings of the same species have no hope of rising to stand with the giants.

Human institutions also have life cycles. A free enterprise system is much like a forest. In its youth free enterprise is an ever changing array of many species of small businesses. The economy brims with the vigor of life. All who share the domain benefits from the youthful vigor. With time a few young trees of the business world rise to overshadow their contemporaries. There is now much less vegetation in the economic forest. Yet, growth remains vigorous.

As time passes the weaker trees die in the shade of their stronger brethren. The survivors dominate the landscape. It is only a matter of time until the giants of the corporate forest cast their deadening shadows over the forest floor stifling new economic life. These aging corporate giants, drained of the vigor of youth, produce no new economic wood. Mainly they stifle the growth of new businesses, which in a healthy economy would spring up to challenge and replace the old, like the young bucks in a herd rise to challenge and dethrone the grey beards.

Maintaining a thriving and productive forest requires cutting the old growth to make way for the new. Old trees, though devoid of growth and filled with rot, are still impressive sights. Their appearance belies their condition. To maintain vigorous life and growth in the forest, those trees which produce nothing but shade must be removed. The same is true in the business world. If free enterprise is to provide more than a brief bust of prosperity followed by decay and decline, we must find ways to cut down the dying corporate giants to make way for the vigorous young businesses that are the only hope for our future.

Eventually the dead and rotting wood in an old forest provides fuel for the fire that makes way for new life. Of course, most other life in the forest also perishes in the flames. Unless we cull out the corporate giants filled with dry rot, in due time we will suffer an economic forest fire that will wipe out the old and make way for the new. The worst thing we can do is convert dying national corporations into multinational ones whose deaths will have even more severe consequences.

Putting giant corporations on government life support, as we now do, will only compound our problems and worsen our fate. If we fail, the only alternative is to await the economic forest fire that will scorch us all.

FREE WILLIE?

Millions among us see zoos as evil. Animals should live free in the wild. Why? Aren't zoo animals well cared for? Don't they receive food, shelter and medical care without lifting a paw? When they grow too old, someone even arranges their transfer to that great zoo in the sky. Why should we turn these favored creatures back into the wild to risk starvation, disease and mutilation? Some people still insist that animals are meant to live free – that risking the perils is the only way those animals can be what they should be. The animals' position on this issue is a bit unclear. I guess Dr. Doolittle forgot to ask.

Insulating people from the perils of life is not doing them a favor. Their only hope for success and satisfaction is to face the perils and conquer them. When we take away the risks, we destroy the opportunities.

Is it not strange that millions of people, often the same people, want humans treated like zoo animals? Should government guarantee all our needs so that we never risk failure or disaster? Should we really treat animals like people, and people like animals?

The animal in the zoo contributes nothing to its own care and keeping. The wild animal does everything for itself. We humans are able to produce enough food for lots of animals. Putting animals on welfare isn't a major problem. We control their breeding too. Thus, a burgeoning population of zoo animals is not generally a problem. What happens though when we put that last person in the human zoo? Who will be left to feed us? Will the wild animals take care of us? – I don't think so.

Besides, what about the joys and wonders of living free? Can only animals appreciate those joys of freedom and risk? I recall an episode of the *Twilight Zone.* A small-time thug had gone to his final reward. He found himself in a casino where he could gamble all he wanted. Soon he discovered that he always won. He was ecstatic at the thought of winning forever. He commented to an attendant that, he never dreamed Heaven would be like this. The attendant responded, "Where did you get the idea that this is Heaven?" Have we lost sight of the fact that success and satisfaction are impossible without risk of failure? If you take away the risk, people will create their own. This may explain why the second generation rich venture into auto racing and the like. It is likely that it also explains some of the risks of death and destruction taken by welfare recipients.

Insulating people from the perils of life is not doing them a favor. Their only hope for success and satisfaction is to face the perils and conquer them. When we take away the risks, we destroy the opportunities. When we create an environment where a person gives up the battle, we have destroyed that person. Not only that, we have set in motion the process likely to visit that destruction onto future generations.

Risk of failure inevitably leads to some failure and the consequences there of. We can extend a hand to those beyond the point of helping themselves. If they will not grasp the hand and pull, there is nothing we can do for them except perpetuate their misery, and allow them to pass it on to their children. At the end of that road lies the destruction of our civilization. Our future doesn't depend on letting animals face the risks of freedom. It does depend on requiring the members of the human species to face the risks of failure and the opportunities for success.

FUN

Everyone likes to have fun, Right? Is there anything wrong with that? Didn't the Founding Fathers enshrine the pursuit of happiness as one of our most sacred rights? Don't fun and happiness grow on the same tree? Aren't most people who are having fun happy? How many people are happy when they aren't having fun? So, why shouldn't we strive to make our lives wall to wall fun? Fun at school, fun at work, fun doing household chores.

There are a couple of flies in the ointment. One person's fun may be another's drudgery. Even those fun things get dull and boring after while. That video game which at first was more important than food, eventually ceases to cast its spell. Millions of people spend billions of dollars visiting Disney Land, Cedar Point, and other amusement parks. Yet, if these same people were forced to spend every day at an amusement park – How long before they would beg for mercy?

The most grievous child abuse practiced today is the failure to prepare children for life in an unfun world.

Few people, if any, find unending joy in any endlessly repeated activity. Yes, that list includes sex. In fact one type of therapy given certain sex offenders is forcing them to watch endless X rated movies until they are totally bored. Try making a list of things that you find to be fun. See how many activities there are on that list which you get to do as often as you want. How many of the things you enjoy the most are the ones you get to do the least? Doing the fun things as often as we want is far more likely to end the fun, than to enhance it. So, those people who seek to make life endless fun face a real challenge.

For most of the history of the human race fun has been on the back burner, if on the stove at all. Those front burners were dedicated to cooking the pot of survival. The first law of fun is that those who don't survive don't have any, at least not in this life. Now that survival is easy, we have rotated the pots. If you don't think survival is easy now, read some history. We now constantly stir the fun pot on the front burner while leaving survival to simmer by itself.

Not that many years ago children grew up in an environment where they learned by endless experience to do what had to be done. Household chores, work in the family business, endless lessons and exercises in school, prepared the child for an adult life where he did what had to be done without the endless lament, "This isn't fun." No one expected life's tasks to be fun. Most did understand that these tasks had to be done.

Herein lies the biggest failure in our society today. Children grow up in a fun oriented world with little work and few household tasks. We even excuse them from learning because of our own failure to make school fun. How are they to be prepared for adulthood that includes all those unfun things that must be done to achieve anything approaching a productive life? The most grievous child abuse practiced today is the failure to prepare children for life in an unfun world.

In a stable successful society, fun of necessity must be the desert, not the main course. With our obsession for making this a fun world for our children, we are doing the equivalent of feeding them endless cake and ice cream. We seek to hide the vegetables in the desert.

We must return the fun pot to the back burner where it belongs. Again, we must prepare children for life in a world of boring tasks that must be done. If we don't, we will continue down the road toward collapse of our civilization. A big part of the problem is that adults raised in a fun worshiping world aren't up to the task. Preparing children for life in an unfun world isn't much fun.

Know Where You Are Going, Before You Run

When lost in the depths of a swamp, Which guide would you choose to follow? One is certain he knows the way out. He wants to run out of the swamp and have everyone follow. The truth is, he only runs in circles. He does sincerely believe he knows the way, and truly wants to get out of the swamp. The second guide admits he doesn't know the trail to the edge of the swamp. He does believe he knows the direction to start. He asks you to join him in a slow search for the right trail. He admits there will be false starts and back tracking. He insists that together we can find the way.

> **We need leaders, not pushers. Solutions must be sold, not force fed. If the federal government wants to be a part of the solution, its leaders must learn to exercise leadership. They must inspire and motivate people across the nation to go forward on their own with solutions that will work.**

The two guides are not unlike the voices that rise across the land asking our allegiance in finding our way out of the social swamp into which our nation has wandered. If the politicians and others were open minded and honest, they would all admit that they don't have the answers. They would acknowledge that there are no magic solutions. No laws or programs that government can implement across the nation will solve the problems of crime, welfare, education, drugs or anything else.

There are some projects that are helping at the local level. To force these onto the entire nation is folly. Not all communities are the same. Even more important, the local projects that work succeed in large part because they are run by people with leadership ability and a deep dedication to making the projects work. The projects that work invariably have enlisted the active participation of many members of the community.

Consider the neighborhood police programs where officers are assigned to a neighborhood to work with the residents as part of the community. Such programs have proven successful. I firmly believe we should have more of them. So, I should support the 1994 crime bill that seeks to implement these programs across the nation. Wrong. Such national legislation is worse than foolish. If the people running the programs don't believe in them, and have the training and skill to make them work, the programs are doomed from the start. A poorly implemented neighborhood police program is worse than nothing. Send the wrong kind of officers to do the wrong things and the program will alienate people, rather than rally them to revive their neighborhoods.

This explains why many programs that start with promise fail miserably on a large scale. We need leaders, not pushers. Solutions must be sold, not force fed. If the federal government wants to be a part of the solution, its leaders must learn to exercise leadership. They must inspire and motivate people across the nation to go forward on their own with solutions that will work. A forced march will not work, even if it starts in the right direction.

More often than not we aren't even marching in the right direction. Nearly all nationwide programs are ill conceived and ill advised. Until the trail is tested and proven to lead in the right direction, it is worse than foolish to demand that we all stampede in that direction. A quick look at the past 60 years shows that most of these "promising" trails lead strait over a cliff.

Education bills, crime bills, drug wars, none have worked. Yet, our so-called leaders still scream for us to join the latest stampede. In politics those who promise the most deliver the least. The promises do serve their purpose. They win elections. We, the voters, must seek out and follow the quiet voices of those who admit they know not the answers – the ones willing and eager to lead the search for those elusive solutions. Until we do, we will continue stampeding down the false trails leading deeper into the swamp. We must elect candidates with open minds and leadership ability. We must reject candidates who claim they have found the answers. They are liars and con artists of the worst kind. Anyone who wants to impose a nationwide program of any kind (Unless the program is a matter of foreign affairs, such as immigration.), is dangerous and should not be trusted. It matters not whether he rides a donkey or an elephant, or is a cross between the two (a short guy with big ears).

Kissin' Cousins

Most of the big battles in Washington and government in general are portrayed as conflicts between liberals and conservatives. Support of either side in the battles over welfare, education, gun control, abortion, or any other key issue, is certain to earn you the brand of either liberal or conservative.

The irony is that most people on both sides of that liberal-conservative fault line seek similar results. All the participants in the great battles (except the politicians and bureaucrats who owe their power and incomes to the misery of others), want people to be self supporting, well educated, and to stop killing, robbing and raping. Most would also prefer to create an environment where there are few abortions. I see little evidence that there are many pro choice people who would recommend to a woman that she become pregnant so that she could partake in the joys of an abortion.

Giant corporations are not the crown jewels of free enterprise. They are the malignant tumors that destroy free enterprise.

If liberals and conservatives are not divided by their goals, Why are they prepared to fight to the death like two cats with their tails tied together and hanging over a clothesline? The conflict between liberals and conservatives is over the road to travel to the goals. Each is certain they know the way and that the other is headed down a dead end street that goes over a cliff.

At least one group has to be wrong. It is at least arguable that both routes could lead to either success or to failure. It is not possible that either route will lead to both complete success and total failure.

The liberals' travel plan follows the expressway of government action. For every problem the liberals see a government program as the road to success. Individual choice, business and free enterprise are the enemies that must be contained. The conservatives seek to travel the road of free enterprise and private action. It is government that must be contained and put in its place before we can progress toward solutions for the problem *du jour*. Still, the conservatives prescribe large doses of government for certain maladies. Who is right? Who is wrong? Both are right, and both are wrong.

It is easy to catalogue the failures and sins of both government and business. If the evil chickens of either come home to roost, the flight will darken the sky. To understand the problems we must seek the reasons for the failures. The universal lesson from history is that those who have power abuse it. Both liberals and conservatives seek to fight power with power. Each see themselves as the champions of good power and the opponent of bad power.

The problem is that success guarantees failure. Once power is achieved, it is only a matter of time until it's abused and becomes the new oppressor. The only real answer is to disburse power and break up the power centers that exist today.

Now government is the foremost oppressor. Not only does government oppress, it has also failed to deliver the good it has promised for decades. Failure cannot be hidden forever. If big corporations are not full-fledged conspirators in this oppression and failure, they are at least willing fellow travelers. Some may see business as the leader and government as the fellow traveler. Whichever way you see it, the result is the same. If we look at history we will find that government, business, churches, and labor unions have all had their turns at the top of the power structure. We will also find that the average citizen didn't fare very well under any of these dominances.

If we look at history we will find that government, business, churches, and labor unions have all had their turns at the top of the power structure. We will also find that the average citizen didn't fare very well under any of these dominances.

In our quest to quash the current government monster we must not ignore the pretenders who will gladly fill any power vacuum left by the shrinking of government. The current crown prince is the multinational corporation that already plots to displace government as the dominant world power. If we diminish the power of government without emasculating the evil crown prince, we only aid his dark cause.

Giant corporations are not the crown jewels of free enterprise. They are the malignant tumors that destroy free enterprise. The vigor and success of free enterprise spring from free markets and competition. The monopolistic powers of corporate behemoths destroy both. When the destruction of the healthy body of free enterprise is complete, all that will remain are the corporate tumors.

The realization that big government is a problem, not a solution, fuels the growth of the power of the conservative forces in the political wars. The faith is misplaced. It is a myth that most conservatives seek to shrink government. Both conservatives and liberals seek to prune from government the branches they despise. Each also seeks to graft on new growth of their choosing. The disagreement is far more over the purpose of government than its size. Most conservatives and liberals are kissin' cousins.

Instead of liberals and conservatives battling to empower their champions, they should join to disperse and decentralize the power of both government and business. This is the only way to protect freedom and to stimulate the individual responsibility and action essential to maintaining a livable world for all of us and our children. Big government is dangerous, no matter what its mission. A world dominated by oppressive corporate giants will be no more pleasant than one dominated by oppressive government. Is the name of the giant who kicks you in the face of great importance?

A Fair Hunt

Anti hunting activism has washed upon us for sometime. At first glance it seemed that most of these anti hunters were some kind of kook. On reflection I believe I am beginning to see their point. Hunting as it is now generally practiced is discriminatory, unfair and downright un-American.

To prove the point we must look at the habits of the average hunter. What do most hunters have in common? They hunt where they are most likely to find. Not only that, they concentrate their efforts during the times of day their quarry is most likely to be out and about. Pheasant and deer hunters concentrate their efforts in the early morning and late afternoon. How many coon hunters stalk their prey during the daytime?

Even worse. Deer hunters are likely to hunt around trees, brush, or other cover. How many deer hunters have you ever seen sitting in the middle of a 30 acre plowed field? Some hunters even concentrate their efforts on places where deer eat and drink. Rabbit hunters are just as bad. They seek the brush and fence rows where rabbits live.

Why shouldn't hunters hunt where they are most likely to find? I don't make the rules. I only report them. Granted, those nine black robed guardians of the people, anointed to the Supreme Court, have yet to fully expound upon the Constitution's scheme for hunting.

Turkey hunters are some of the worst offenders. They don't even bother hunting in the parts of the country where there are few or no turkeys. Deer hunters have another nasty habit. They are very selective about what they shoot. If the deer they see is not the one they seek, they are likely to let it pass. Fishermen are more fair minded. They catch everything and then sort them out.

Some may ask, Why shouldn't hunters hunt where they are most likely to find? I don't make the rules. I only report them. Granted, those nine black robed guardians of the people, anointed to the Supreme Court, have yet to fully expound upon the Constitution's scheme for hunting. What they have said leaves little doubt how they will view other hunters when those worthy Justices find time to impose their will on the rest of the world of hunting.

We do know what those Justices, and the lesser lights in the lower federal courts, have said about hunting for drunk drivers. If a game hunter went after drunk drivers, the hunter would likely concentrate his efforts in the vicinity of taverns and parties during the late evening and night. He might allow drivers who didn't appear drunk to pass. How un-American. The courts have made it clear they will have no part of such discriminatory conduct. Woe be to the hunter of drunk drivers who focuses on the prime habitat, especially in the prime time of day (or night). Not only that, the courts frown upon stopping only the drivers most likely to be drunk. The hunter must catch them all and throw back those who aren't keepers. Anything else would be unfair discrimination.

Systematic stopping of drivers is okay if the area for making the stops is selected at random. The officer must not use any judgment about which might be drunk. He must stop them all. How many drunk drivers are likely to be bagged near a church Sunday morning, unless someone has really spiked the communion wine? What about drivers leaving a tavern in the wee hours of the morn? Still the rules say the hunter must take leave of common sense and ignore the obvious.

Sure, some unenlightened souls may argue that the courts are backward and the hunters are on the right track. How dare we question the wisdom of our elite guardians wearing those black robes we are not even worthy to touch? Is it possible that black robes affect brain cells the way kriptonite affects Superman? (You are instructed to disregard that last remark.) All hunters must adopt the same principles of fairness. So what if the hunt is less successful? 'Tis but a small sacrifice for fairness.

Even equality has its limits. My trap still sets by the woodchuck hole. An exception for worthy trapping does not necessitate one for hunting. How did I discover the basis for my exception? Compare the rules for searching for firearms to those governing searching for drunk drivers. But, aren't firearms a great hazard? That they are. If the trend continues, murders with firearms may someday kill as many Americans as are now routinely killed by drunk drivers.

THINK BIG

Are you lonely and depressed? Worried about how you will pay the heating bills next winter? Fed up with both political parties? Do you despair that you will never live in a land where everybody knows your name? Cheer up. Help is on the way. Success beyond your wildest dreams is no further away than your mail box. All you need do is harness the power of junk mail.

Not long ago I ordered books from two different book sellers. These innocent acts lifted me from obscurity to the prime sucker list. The blitz of book catalogues landing in my mail box may well affect the rotation of the earth. With the magic of mailing lists, book sellers I never heard of now count me among their dearest friends. How do I know? They are all writing to tell me so. Already the catalogues I received far exceed the weight of the books I purchased. And the influx continues unabated. In fact, the flow appears not even to have peaked.

Your name will also be embossed on the Golden Fleece mailing list and distributed to all like minded political causes – and a whole bunch of totally mindless ones.

With a little imagination I can attract enough junk mail to heat my home for an entire winter. There will be enough left over to fill the pot holes made by the overloaded mail trucks that deliver my bonanza.

Some people use chain letters in desperate attempts to generate a flow of mail into their mail boxes and the mail boxes of others. How unimaginative. Just check the "I want more information" box on the postage paid cards sent to you, or tucked away in magazines. You can be everybody's pen pal without spending a dime.

What about those political parties? Doing them in will cost some money. Ten dollars per person per party should nicely do the job. Don't send the ten dollars to me. I'm not forming the Hit Men's Benevolent Association. Send the $10 to the national committee of the party of your choice, and get your friends to do the same.

You may ask, How will sending money to a political party put it out of business? Such questions only reveal how sheltered a life you lead. Giving mailing lists to politicians is like giving whiskey to alcoholics. That $10.00 contribution entitles you to membership in the solicitation of the week club. But wait – there's more. Your name will also be embossed on the Golden Fleece mailing list and distributed to all like minded political causes – and a whole bunch of totally mindless ones.

Your mail box runneth over. The printing and postage costs for those letters will quickly eclipse your $10 contribution. Not only that, "friends" you never knew will start calling you to make sure you are alive and well – and to be sure you understand the urgency of their needs. The compulsive junk mailers will bankrupt the party.

If everyone sends in that $10.00, the forests will disappear from North America to feed the printing presses. You will know you have made a difference when those letters start arriving with the legend at the bottom proclaiming, "Printed on Spotted Owl skin." You may say, "What's so great about exterminating forests and spotted owls?" Quit complaining. Do you want to get rid of the political parties, or don't you? I never said there wouldn't be side effects. What do you expect for $10.00?

TREATMENT

In spite of all the advances of medical science, cancer treatment is still devastating and often dangerous. Few would be likely to volunteer for cancer treatment, except that the alternative is usually suffering and death. Faced with the alternatives, many of cancer's victims consider drastic measures, such as surgery, radiation and devastating chemicals, to be the lesser of the evils. Some choose to deny the seriousness of their condition. Others turn to quacks who promise simple, easy cures. I'm not saying that all those who propose alternative treatment for cancer are necessarily quacks and frauds; certainly many are. Remember a few years back when hundreds of people trekked to the Philippines for bare hand surgery?

Denying the existence of cancer may bring peace of mind for a time. Delaying treatment only increases the certainty of the end. A treatment that only slows the growth of a cancer may be of some benefit. Still, it only postpones the inevitable. We don't hear much about reforming cancer. The elusive goal is to wipe it out. Attempts to minimize the ravages of cancer are seen only as stopgap actions along the road we all hope leads to eradication.

A social malignancy runs amuck in our land today. If left unchecked it will yield far more human suffering and misery than all the physical cancers ever inflicted on humankind.

A social malignancy runs amuck in our land today. If left unchecked it will yield far more human suffering and misery than all the physical cancers ever inflicted on humankind. Yet many, if not most, of our fellow citizens are still in the denial stage. They fear the treatment more than the disease. The mere fact that this malignancy has not yet destroyed us is no reason for comfort. The same as with physical cancer, the longer we live with it, the closer we are to the end. Also just like with physical cancer, slowing the rate of growth only postpones the inevitable a bit longer. A social cancer may take decades, rather than years, to exact its price. This doesn't make the end any less certain.

The social cancer of a big, smothering government clogs the arteries and destroys the organs of the society it infects. We should take no comfort from the fact that our society has the capacity to survive the spreading malignancy for a time. *U.S. News* and others have reported that by about 2012 interest and transfer payments (money paid to people simply because they exist) by the federal government will consume all revenue produced by taxes now in effect.

If the citizens of our country don't act quickly to treat the malignancy, there is only one possible end. How much comfort should we take that the end may be a few decades away? Is it victory to postpone the inevitable demise of our society so only our children and grandchildren are left alive to suffer?

If the citizens of our country don't act quickly to treat the malignancy, there is only one possible end. How much comfort should we take that the end may be a few decades away? Is it victory to postpone the inevitable demise of our society so only our children and grandchildren are left alive to suffer?

Treating and defeating the social malignancy of gargantuan government will be no less traumatic and no less risky than treating a physical cancer. The only alternative is to accept the inevitability of the end and go quietly. Has big government already so dispirited the people of America that they are willing give up without a fight? Perhaps there are still too many people in the denial stage. If that is the case, we best start soaking them with reality. The malignancy does not yet appear to be inevitably terminal. A decade from now that may no longer be true. One thing is certain, the longer we wait to start the treatment, the more difficult and traumatic it will be.

DO YOU WANT TO BE MURPHIED?

Are there any among us who have not heard of "Murphy's Law?" That time honored principle proclaiming, "Anything that can go wrong, will go wrong," is now a legend. Is it a valid principle? That depends on how you see it. Certainly everything that can go wrong doesn't go wrong every chance it gets. On the other hand, "anything that can go wrong, will go wrong eventually," is a truism that should stand without question.

If you drive enough automobiles, down enough roads, enough times, every possible accident will happen. If there are enough Lotto drawings every possible number combination will win. Stick with your favorite number, it's bound to win eventually. So what if it takes 50,000,000 drawings. Patience is a virtue.

The Founding Fathers were scared to death of that little biddy federal government they created. That's why we have the Bill of Rights.

Murphy's law also applies to government. Any government strong enough to take away your freedom, will take away your freedom. It isn't a question of "If?," only a question of "When?"

The Founding Fathers were scared to death of that little biddy federal government they created. That's why we have the Bill of Rights. The Bill of Rights protect the states, and their citizens, from the federal government. More than a half century passed before the Supreme Court came up with the idea of using the Bill of Rights to limit action by the states.

Today one regional office of one department of the federal government spends more money and has more power than the entire federal government did in George Washington's day. In case you haven't noticed, the federal government has many departments and many regional offices. If that doesn't scare you, it should. When will the power be abused? It would take more than a book to document the flagrant abuses already inflicted.

Some people think we have free speech. How free is it when the federal enforcers threaten prosecution, confiscation and ruin to those whose only sin is speaking out against a bureaucrat's pet project? Obscenities, and often slander, enjoy the protection of the Constitution. Speaking out against government does not, at least not in the minds of bureaucrats. The Department of Housing and Urban Development threatened prison and ruin for those whose only transgression was speaking out against a HUD project. Maybe I'm missing something, but isn't that the opposite of what the architects of the constitution intended? Wasn't their primary interest in free speech, free press and all that sort of stuff, to limit the power and abuse by government? Not just to let a few freaks be as vulgar and disrespectful as they can be.

Living in a country with a large and powerful government is like living in a cage with a hungry tiger. It isn't a question of if it will eat you, only when. Millions of Americans take far too much comfort from the fact that so far the tiger eats only other people. Many appeasers are even willing to feed their neighbors to the tiger to keep it at bay. They usually feed the neighbors they don't like to the tiger. When the tiger calls at their house, those appeasers will wish they still had some neighbors to call for help.

One of the tiger's latest feeding frenzies is directed at smokers. I don't question that smoking would be a good riddance. Not all agree with me. Do I have the right to impose my values on others with the force of government? If I do, they have the right to do the same to me. When the smoker blows smoke in your face, you should have the right to object, and to prevent the assault upon your person.

If you want to bring social pressure against smokers to get them to stop, that's okay too. Effective use of pier pressure is the keystone of civilization. You should also be free, as an individual, to favor or disfavor others for eating meat, not eating meat, growing dandelions, not growing dandelions, wearing green and white or maize and blue, and a million other things. We should not for a minute tolerate an all powerful government using that power to further any of these pet causes, whether they are our own or someone else's. If we allow government to do these things, or even allow it to have the power so it can, it is only a matter of time until we get Murphied too.

MYTHS

Where have all the myths gone? When we look back on history, we find myths from all the ages. How is it we were fortunate enough be born in the age of enlightenment, the first generation of humankind to rise above myths?

Sacrifices to the mythological gods of the Greeks and Romans continued until the worshipers recognized that the gods were myths. Our foolish and equally useless sacrifices to the gods of education will also continue until we the people see through the myth that money can lift us to educational heaven.

What is a myth? "A fiction or half-truth, especially one that forms part of the ideology of a society . . .," so says the *American Heritage Dictionary*. Myths are false. Yet, we believe they are true. The mere fact we know not that we have myths does not mean we are myth free. Myths are like spies, to identify them is to destroy them. The only myths we can recognize are those that belong to the past, or to someone else.

Those past generations were as certain of their freedom from myths as are we. To indulge ourselves in the belief we have no myths is as dangerous as assuming there are no spies whom we haven't caught. In times of international crisis, survival depends on eternal vigilance for spies. In times of social crises, our future depends on how good a job we do unearthing and debunking our myths. The head in the sand approach may bring short term comfort. It doesn't aid our survival.

Space does not permit me to wrestle with all the myths ingrained in our society. It would be presumptuous for me to even suggest that I have somehow unmasked all of our myths. For now I will content myself with a little tug on the mask of one myth that already is starting to blow its cover.

60 Minutes ran a segment about the Kansas City, Missouri school system. Years ago a federal judge ordered Missouri to open up its

coffers for the benefit of the Kansas City schools. The money ran deeper than a flood on the Mississippi. Well over a billion dollars flowed for improved facilities. They spend $44,000,000 a year on transportation alone. It really is "transportation alone." Some students ride to school in taxi cabs. The facilities are impressive. Computers sprung up in the classrooms like mushrooms. Teachers' salaries rose. Class sizes dropped.

The Kansas City schools did everything good schools should do, except one minor detail. The great wash of money didn't improve the performance of the students. Achievement scores remained unchanged or even dropped a bit. Students in one school did buck the trend. Why? Was it because that school had admission standards? Was it because the parents were required to agree to get the kids to school on time and to make sure the kids did their homework?

A while back a national columnist looked at education in America. He cited a study of results verses expenditures. The states that were getting the best results were more often that not the ones that spent the least money per student. The column pointed out that quality of education in America bore more relationship to proximity to the Canadian border than to money spent. Maybe we should move all states north. This approach offers little hope for Michigan. Those of us who didn't flunk geography know Michigan is pretty well nestled up to Canada already.

The idea that more money, and only more money, equals better education is one of our dangerous myths. This myth is continually used to bludgeon anyone who would deny a request for money for education. Anyone who opposes spending on education is branded as against children. Who dares to be against children? Those whose livelihood depends on the flow of money to education spend millions on slick add campaigns to support the myth. Sacrifices to the mythological gods of the Greeks and Romans continued until the worshipers recognized that the gods were myths. Our foolish and equally useless sacrifices to the gods of education will also continue until we the people see through the myth that money can lift us to educational heaven.

Education requires money. The flow is already more than adequate. Look at the last 50 years. The more we spend the less we get. When I started school, my school spent less than $150 per student per year. Now $5000 is inadequate in the eyes of many people. There has been some inflation, not 3300 percent. It's the students from those $150 schools that brought strength and prosperity to America. What will the $5,000 students do? Will they be better myth busters than are we? Myths do not die easily. To the believers, a myth is a sacred truth. Throughout history millions have died in defense of their myths.

A dedicated physician battled for over ten years to destroy the myth that stress causes stomach ulcers. He overcame the scorn and ridicule of his "learned" colleagues to bring us the truth that ulcers are mostly caused by bacteria. When "scientific" myths are this deeply rooted, What will it take to kill a social myth?

WHO WILL RAISE THE CHILDREN?

During the last century and the early part of this one, millions of immigrants set sail for America. Though these immigrants sought a better life, most didn't rise above what today would be considered extreme poverty. Does this mean that these immigrants lost when they gambled on America? No way! There is no absolute measure of success. Success is measured by comparing accomplishments to the goals sought. These immigrants found the opportunity to earn a living for themselves and their families. They raised and educated their children to be able to rise to a higher level of security than the parents. This was success. Until the most recent decades, few people achieved more than this in life. Few expected more. I still have memories of a now long departed generation that shared these same modest expectations.

When the young adult discovers that the world doesn't meet his or her demands as quickly as the parents did, the adult sees the problem as being with the world and demands that government (the new adopted parent for adults) fix it. Children run to mommy or daddy. Adults run to government.

People focus their energies on those things they find most important. Those people who saw success in life as raising their children to be productive, secure adults, left no stone unturned to achieve that goal.

The increased productivity of the late twentieth century pushed the concern for survival to a back burner. Unfortunately the concern for the raising of children went along with it. Parents are still concerned about their children. There are many other concerns that compete for the parents' attention. Parents see many other roads to what they call success. Leisure, thrills and pleasure absorb much of the attention once given to children. In our materialistic society, parents are often more concerned about the physical things they can provide for their children, than about the values and habits that prepare the children for life on their own.

Even the more responsible parents commonly turn their children over to juvenile baby sitters and television for extended periods. At the same time they shower the children with the latest and most expensive gadgetry. The parents raise the expectations of the children without teaching patience and productivity. The child that never has to wait for the new bicycle or video game grows to adulthood expecting instant gratification. When the young adult discovers that the world doesn't meet his or her demands as quickly as the parents did, the adult sees the problem as being with the world and demands that government (the new adopted parent for adults) fix it. Children run to mommy or daddy. Adults run to government.

Downgrading the importance of raising children is the greatest evil of our times. For the survival of civilization, training children is the most important job anyone can do.

All of the social problems (crime, poverty, poor education, drugs, etc.) are byproducts of the way we raise children. No society has ever done a perfect job raising children. Each generation in doing less of a job than its predecessor. Unless we reverse this trend our civilization is doomed. It isn't a question of "If?," only, "When?"

I am not suggesting that parents must devote all their energies to raising children. Parents must devote more energy to the instruction and training of children. Of course, unless these energies are directed in the right direction, there will be no improvement. There is no substitute for a constant, responsible adult influence in the lives of children. It is beyond the scope of these comments to explore what we can do about parents whose upbringing and education leave them totally incapable of training and instructing children. This is, however, a problem we must solve.

The way we raise children is largely a product of the environment in which we live. That environment changed radically during this century. The nineteenth century parents did a far better job than we are in spite of (or perhaps because of) having far less economic opportunity than we do. Claims of lack of opportunity are copouts.

Even those in poverty have far more opportunity than the citizens of the nineteenth century. They just don't know how to recognize and use those opportunities.

Downgrading the importance of raising children is the greatest evil of our times. For the survival of civilization, training children is the most important job anyone can do. The careers we pursue, the wealth we accumulate, the thrills we seek, all will mean nothing 100 years from now. The way we raise our children will determine whether civilization still exists, and what that civilization will be.

Supposedly Nero fiddled while Rome burned. Rome was only a city. We are fiddling while our civilization burns. Will history treat us more kindly than it has Nero? Is there any reason why it should?

Is Price More Important Than Quality?

Buying a car requires some important decisions. Making the wrong decisions can be costly. It could ruin the proud owner's whole day to discover he could have saved $2,000 by driving across town to another dealer. It could ruin a lot more than just a day to discover that the new chariot is a hopeless lemon. If the vehicle is so useless that it will cost more than it's worth to fix it, the buyer hasn't just lost a few dollars, he has lost every cent he paid. Hang on, it can get worse. The lemon could be dangerous. The buyer or his family could be seriously injured. Many things can go wrong when you buy a car. Paying a few dollars too much for a good vehicle is one of the smaller risks.

The mean spirited people who don't care aren't the ones who object to cost, they are the ones who promote useless and dangerous programs in the name of trying to help.

Car buying time isn't the only occasion when we face momentous decisions about cost and quality. Before every election the politicians trot out their wares. They tell us the wondrous things the new models of government programs will do for us. They strive mightily to make those new models look different and enticing. Just like when buying a car, it's best to check what's under the hood, as well as the price tag.

Most objections voiced about government and government programs focus on the cost. Whether it's education, welfare, health care, crime control, or something else, the big question is, "What will it cost?" The most vocal objections are, "We can't afford it."

We spend far too much time barking up the cost tree. The real question should be, "Is the merchandise the politicians are peddling any good?" If the grandiose programs attempted and proposed by the federal government would do even half the good their sponsors claim, they would be bargains at twice the price.

Attacking government programs on the grounds that they cost too much leaves the attackers open to charges that they are greedy and just don't care. The federal government has been pushing social programs since before I was born. None of them have solved any problem without creating a worse one to take its place. Just like with buying a car, we should focus on the quality of the merchandise, not the price.

Even the total failure of the programs isn't their worst feature. The worst is that such programs almost always leave us worse off than when we started. They are like the defective car that causes serious injury. The mean spirited people who don't care aren't the ones who object to cost, they are the ones who promote useless and dangerous programs in the name of trying to help.

Space does not permit even a superficial look at why all massive social programs are doomed to failure. That is a book in itself. Doesn't the fact government has over 50 years of continuous failure in the social engineering arena, give adequate reason to suspect it isn't about to put together a winning season?

It's well past time for an all out assault on government social programs because of their failure, not because of their cost. Until a majority of the voters recognize that solutions will never come from Washington, we have no hope of stopping the erosion of the quality of life in America. It's far better to walk than to wait for a bus that will never come. It's even worse to get on the government bus because it's going in the wrong direction.

The Struggle Into Slavery

While Socrates was in prison near the end of his life, his only possession was one earthen bowl from which he ate his meager meals. While Socrates washed the bowl it struck a rock and shattered. Reportedly Socrates exclaimed, "Free, free at last!"

In a twentieth century corollary a dairy farmer explained the ownership of cattle. He asserted that it is incorrect to ask, "How many cows do you own?" The correct question is, "How many cows own you?" Anyone who has tended a dairy heard needs no explanation of this truism.

Throughout history, and even today, most of the world's population had no reason to worry about being enslaved by possessions. Eking out enough to survive for a few more days or months is the far greater concern.

Like a great castle, we build our houses of plenty one brick at a time. Unfortunately we focus our gaze on the bricks not laid, rather than on the ones already in place.

The wealth of ordinary citizens in the industrialized nations exceeds that of the nobility of the not so distant past. We didn't fall instantly into this wealth. Like a great castle, we build our houses of plenty one brick at a time. Unfortunately we focus our gaze on the bricks not laid, rather than on the ones already in place. Thus the common lament, "If only I had a little bit more, everything would be okay."

How is it that awash in the wealth of the most productive era in human history, surveys show that most young Americans despair of ever achieving "The American Dream?" Might I be so blasphemous as to suggest that the problem is with the dream, rather than with lack of income?

What is this "American Dream?" The media defines it for us. The most common definition is that each generation must rise above its predecessor. Simply stated children are supposed to have more than their parents. Not only that, most children expect to get their life time achievement awards about three weeks after they start work.

They want now what it took their parents years to earn. I'm wrong on that point. They want more. The younger generation spends billions of dollars on gadgets and devices that many members of the older generation ignore. Teenagers now spend more on nonnecessities than adults did a few decades ago.

The dream of the younger generation is to have everything they want and still achieve the level of financial security and leisure time their parents enjoy. Never mind that the parents have put the cost and time of raising children behind them and often have small or nonexistent mortgage payments.

The younger generation does its best to achieve its dream. This means borrow and charge. Constantly living one pay check from financial disaster brings on despair. Miss a pay check and there is no money for food. Besides, foreclosure and eviction loom ahead. Why is there such a problem providing the basic necessities in the midst of plenty? Mainly for the same reason that working alcoholics and drug addicts have problems finding money for food, clothing and shelter – they spend the money on something else.

We are slaves to our life style – it is truly a fearsome master.

I fear that many people don't realize the amount they spend on non necessities. Microwaves, stereos, VCRs, and a host of other things that didn't even exist a few decades ago are seen as necessities. The list of non necessities includes fast food restaurants, TV dinners and most of the clothing we buy. The food may be necessary. The preparation costs aren't – at least not for those who learned how to cook. Most of the money we spend on food, clothing and shelter is not spent on necessities. When obligated to pay off stereos, wide screen TVs, expensive cars, vacations and gadgets too numerous to list, there isn't much left for anything else.

When our private financial houses of cards tumble, the necessities are swept away with everything else. Millions of people earning good incomes are having real problems making ends meet and paying for necessities. For most of these people, a little more money wouldn't help. They would use the money to make a down payment on another chunk of the American Dream, thus burrowing in a little deeper. The reason they can't make ends meet is they are holding the ends too far apart. We are slaves to our life style – it is truly a fearsome master.

We demand that government become financially responsible – and it should. Individuals should practice some of that financial restrain. Most people want government to spend less so individuals can spend more. Another way of putting it is, if government is more responsible, we can be less responsible. We are barking up the wrong tree. Government will never be more responsible than the citizens. The creation won't rise above its creator.

Our attempts to possess ever increasing amounts of "conveniences" and gadgetry of modern life, destroy our ability to provide necessities without double incomes and double jobs. We are doing it to ourselves. We are the only ones who can solve that problem. If we fail, should we expect sympathy cards from Bangladesh and Albania?

WHICH WAY SHOULD WE GO?

Americans enjoy travel. At least we travel a lot. Maybe we only like to go places. There is a difference. If all you want is to travel, just jump on your horse and ride off in any direction. If you prefer to go to a particular place, a few maps and a bit of planning might be in order. Granted, if you want to go to Memphis and just start driving you will eventually wander into Memphis – if you live long enough. Before you get there you may also wander through Dallas, Boston, Denver and innumerable points in between. If all you want to do is travel, the side trips are just fine. If you see travel as a necessary evil to reach your destination, you may not be so pleased with endless wandering.

During the 1960s a group of political travelers with a faulty compass mounted the Great Society and galloped off on their quest for utopia.

Let's imagine that I'm in St. Louis seeking to reach Minneapolis. I cast my lot with three other travelers who earnestly desire to reach the same destination. After a few hours on the expressway I note with concern that the signs keep mentioning New Orleans. Not only that, the mileages on the signs keep getting smaller. I nudge the driver and express my concern. His response is not what I expected.

"Quit bothering me, don't you want to get to Minneapolis?"

But I think we are headed toward New Orleans.

"Keep quiet. You're bothering my driving."

Let's pull over and talk about this.

"There's nothing to talk about. If you don't want to go to Minneapolis, just say so. I'll let you out."

The driver sealed his lips and pressed the accelerator. I reached over and turned off the ignition and broke off the key.

"You _ _ _ _ _ _! I knew you didn't want to go to Minneapolis. You just want to keep the rest of us from getting there. You're using us to get you to Miami."

Which did the most to further the cause of reaching Minneapolis – Stopping the car or stepping on the accelerator?

Which Way Should We Go?

During the 1960s a group of political travelers with a faulty compass mounted the Great Society and galloped off on their quest for utopia. When anyone questions the direction of that journey the shrill reply is, "You don't want to go to utopia." Those who question the direction traveled by government programs are accused of being mean spirited and against the worthy goals.

Those who oppose the welfare state are branded as against the poor. Those who want to change the direction of school lunch programs must want to starve children. Those who oppose draconian gun control must be for crime. Those who oppose the morass of federal laws directed at race relations are branded as racists. Those who point out we are not headed toward Utopia are accused of not wanting to go to Utopia at all.

Until we stop the out of control driver headed in the wrong direction, we can't even start toward the goals of a saner, more responsible America. We must ignore the shrieks and screams of those who try to prevent us from stopping the political car and unfolding the social road map. Pushing the accelerator is a problem, not a solution, when we're headed in the wrong direction.

Who benefits from increasing poverty in the name of helping the poor? Who benefits by increasing racial animosity in the name of pursuing equality? Who benefits from destroying our schools in the name of education? Who benefits from creating slums in the name of low income housing? Who benefits from child care programs that destroy families and the environment for effective child rearing? Is it so terrible that we ask the drivers of our political institutions to pull into a gas station and ask for directions?

WHERE THERE'S A WILL

A few years ago I attended a seminar in Orlando. While there, I tuned in the local TV channel. The big sports news of the night was, the Tampa City Council had voted against funding a hockey arena to bring a National Hockey League team to the city. What a sad day, no hockey for Tampa. But, wait, Tampa now has an NHL team. What happened? Somebody raised private money to provide the arena. It didn't hurt to ask first. It might have worked.

Most government subsidies are sold to reluctant taxpayers on the basis that the activity is important, and that it will not happen without the subsidy. It's a lie, folks.

Unless you spent the summer at Vermillion, you probably heard something about the much ballyhooed report on a national sex study. Did you also notice that a few years back there was a big campaign to get the federal government to fund this study? The Bush administration said "No thank you" to the opportunity. Once again somebody raised the money privately and did it anyway.

Hockey in Tampa and sex studies are not on my list of the top ten needs in our world. Somebody, though, found both important enough to put their money behind them when government said "No."

In the earlier days of our nation the federal government granted subsidies to get railroads built into the west. Were these subsidies necessary? Without those subsidies would everything west of Pittsburgh still be a vast wilderness inhabited only by Indian tribes and a few trappers served by stage coaches and wagon trains? For some reason, I doubt it.

If denying government subsidies really stops the subsidized activity, we could bring down the tobacco industry, and end smoking, simply by cutting off subsidies to tobacco farmers. Is that why we subsidize tobacco production? Is it the only way to keep those cigarette taxes rolling in?

Where There's a Will

Most government subsidies are sold to reluctant taxpayers on the basis that the activity is important, and that it will not happen without the subsidy. It's a lie, folks. When people really want something done, they find a way to do it. If private funds cannot be raised, chances are that either the promoters of the project don't find their own scheme all that important, or no one else believes in it. In other words the taxpayers are the last resort for funding poorly managed and flaky projects.

The way we now play the game, taxpayers are the first resort for most projects. Just like teenagers will gladly spend their parents' money before spending their own, businesses, "public interest" groups, and all the rest will gladly spend the taxpayers' money first.

You may have noticed, we have a limited supply of money, but no limit on the number or magnitude of projects where that money can be spent. When it comes to choosing between worthwhile projects and boondoggles, the marketplace does a far better job than bureaucrats. Never hesitate to say "no" to government funding. If the project is worthwhile the odds are it will happen anyway. If it isn't worthwhile, Why should it happen at all? Besides, people who put up their own money try harder and do better jobs. They get the job done cheaper too. Government funding can destroy what otherwise might have been a successful effort.

There are a few activities that deserve government funding. Don't be an easy sell. Make the sponsors prove their case beyond a reasonable doubt. It is well past time to stop paying taxes to support projects that either would have happened anyway, or shouldn't have happened at all.

AN OUNCE OF PREVENTION

Fire safety should be on everyone's mind. Fire can breakout anytime, almost anywhere – in the home, the car, the boat, the great out-of-doors. About the only people who can count themselves safe from fire are SCUBA divers. We station fire fighters across the land to respond at the first sight of flame or smoke. When the professionals arrive on the scene, the flames are often as not out of control. Fire fighters are our last line of defense, not our first.

The best way to avoid destruction and death from fires is to prevent fires. The next best thing is to put them out quickly. We teach fire safety in the home, the school and the work place. Still experience tells us, some fires are inevitable. We equip the offices, stores and factories with fire hoses. Many keep fire extinguishes in their homes and vehicles. Insurance companies often require that fire fighting equipment be maintained on the insured premises. Laws require that larger motor boats be equipped with fire extinguishes, even though those boats are surrounded by water. We are all expected to fight fire wherever it occurs.

We pass laws to prevent equipping the premises with crime extinguishers. We tell people not to use a crime extinguisher, even if they have one.

Only the foolish would suggest that on spotting a fire we should all run and hide while we wait for the fire fighters. We the citizens are the first line of defense against fire. Over the years I have extinguished several blazes so small that I soon all but forgot them. I also helped to battle down some grass fires where the issue was in doubt for a time. One thing all these small fires had in common was that if left alone the fires would no longer have been small when the professional fire fighters arrived.

Around the time I started school a grass fire broke out down the road. In less than a minute my six year old friend and I filled an old boiler with water, loaded it onto my wagon, and headed for the fire. For some reason our parents did not allow us to complete our mission. The incident illustrates the instinctive, and quite effective, way we respond to fires in our neighborhoods.

When fires strike occupied buildings both friends and strangers attempt rescues at great risk to themselves. In the process these volunteers save many would be victims. Occasionally a rescuer dies in the effort. Still, we view the rescuers as heros rather than fools. How much would the destruction and death from fire increase if we all dialed 911 and then stood aside to await the arrival of the professional fire fighters?

Yet while encouraging our citizens to fight fire, we do just the opposite with crime. We are taught to call 911 and then hide. We pass laws to prevent equipping the premises with crime extinguishers. We tell people not to use a crime extinguisher, even if they have one. We make it illegal for citizens to carry crime extinguishes in their cars. We even prosecute those who extinguish crime.

Until we learn to treat crime like fire, it will continue to spread and burn. Instead of discouraging and preventing the citizens – our only effective first line of defense – from fighting crime, we must encourage the efforts.

Police are even less effective at putting out crime than fire departments are at extinguishing fires. Crime burns much faster than fire. Usually when the police arrive all that is left of the crime is its ash. The police sift those ashes and file a report duly noting the victim was burned.

Until we learn to treat crime like fire, it will continue to spread and burn. Instead of discouraging and preventing the citizens – our only effective first line of defense – from fighting crime, we must encourage the efforts. We must encourage citizens to equip themselves with crime extinguishers and know how to use them. The would be victim and those nearby are usually the only ones who ever get the chance to extinguish the flash fire of crime while it is still only a spark.

Another thing we do in our battle against fire is make sure we have trained fire fighters and equipment close by. The community too small to afford a full time fire department relies on trained and equipped volunteers. They may not be quite as effective as full time professionals. Still they do a good job. They are far more effective than the best of professionals that don't arrive on the scene until an hour after the alarm sounds.

If police are to be at all effective in preventing crime, we must take the same approach. Trained crime fighters must never be more than a few short minutes away. The communities and neighborhoods that are remote from the police stations must be staffed with trained volunteers who can respond at least as quickly as volunteer fire departments. The volunteer crime fighters should be able to respond even quicker than the fire fighters. For fire fighters to be effective they must arrive with large and expensive equipment. This equipment must be stored at some central location. Each volunteer crime fighter could be fully equipped to respond from his home, business or car. The response time could make the best of fire departments look slow.

Sure some citizen-crime fighters will be injured and even killed. People are injured and even die providing us with swimming pools and bubble gum. There is no totally safe human endeavor. If we are to discourage all endeavors that come with a risk, we must discourage all human activity. With no human activity we would all starve to death in a matter of weeks. Inaction is far riskier than action. Only when we all make crime fighting as much a part of our business as fire fighting, will the flames of crime be cooled.

WHAT IS THE ROLE OF GOVERNMENT?

Most of us recognize that the job of the military is to break things and kill people. In an ideal world we wouldn't spend billions of dollars training and equipping soldiers to smash and kill. In our imperfect world, smashing and killing, and threatening to smash and kill, are the only ways we have to keep hostile nations from doing dastardly deeds, including breaking our things and killing us.

Who would claim that necessity makes breaking and killing good? At best they are necessary evils. When a necessary evil becomes unnecessary, all that remains is evil. We must always exercise great diligence to assure that military action is truly necessary.

Civilian government also has a role in the world. Our vision of that role is less clear than our vision of the military's role. Is there a simple definition for the role of government? What distinguishes legitimate activities of government from illegitimate?

Some attempt to define the role of government based on tradition. Government, for so long as runneth the memory of man, provided fire protection, mail service, roads, courts, schools, police, trash pickup, sewers, water supply, parks, and the like. Private citizens run stores, build factories, raise food, transport goods, build houses, and provide most of the other goods and services that support our lives. Does any of this help define government? No. Both government and the private sector do all of the things on both lists.

Private fire protection is not common, still it is not unheard of – and it does work. Private mail deliveries are now common place. Roads moved in the opposite direction. Though private roads for public use are now rare, in the early days of our nation private toll roads were not unusual. Private schools have always coexisted with the public ones. Private detectives and security guards regularly perform tasks usually considered police work. Many people turn to private arbitrators, rather than the courts, for dispute resolution. In many communities government picks up the trash. In others, private haulers do the same job. Though water and sewer systems are usually provided by government, there are exceptions. Government runs many parks. Still, private recreation facilities are common.

If government isn't defined by what it does, How can we distinguish government from the private sector? Does the distinction matter? It does matter. One of the most troubling issues of our time is defining the role of government. How are we to define that role if we cannot distinguish government from private activities?

One defining characteristic permeates all government activities like a bad gene. That characteristic is coercion. Private ventures expand and succeed by rewarding those who participate. Government coerces us to participate. If everyone is willing, there is no need to involve government. Coercion is the only extra dimension government brings to any activity. The bigger the government, the more it coerces the citizens. The only reason to have any function performed by government is that some citizens will not support that function, unless forced. In a few areas, such as criminal law, force is indispensable. Where force is indispensable, government is indispensable. In all other areas government is unnecessary, and usually counter productive.

One defining characteristic permeates all government activities like a bad gene. That characteristic is coercion.

A private sewer system works just fine, if people want it and are willing to pay for it. Government builds sewers to coerce the unwilling to accept the system and pay for it. Landowners and taxpayers who don't need the system are forced to pay large sums to build the sewers they don't use. Why? Because those who want the system are unwilling to pay for it. If individuals take a neighbor's money to build a sewer, we call it theft. When government does it, we call it public works.

Analyze the functions of government and we find that most could be performed by the private sector – if people were willing to engage in the activity and pay for it. This explains the failure and unpopularity of government. If an activity isn't questionable, government rarely needs to get involved. The activity would go forward without the government push. When government forces citizens to engage in, and pay for, activities which they do not support, those citizens are bound to be resentful and feel abused.

Government motivates with the stick. The private sector uses the carrot. The stick only prompts the minimum effort needed to avoid bad consequences. The stick never generates enthusiasm or excellence. It doesn't make true believers. It only elicits grudging compliance. Thus, government struggles endlessly to achieve mediocrity, and usually falls short of even this modest goal. Only the possibility of reward inspires enthusiasm and greatness. The rewards are not necessarily financial. For practical reasons alone, the only tasks we should doom to the grey world of government are those which are clearly necessary – and impossible without government coercion.

If individuals take a neighbor's money to build a sewer, we call it theft. When government does it, we call it public works.

Most people find some activities of government to their liking. They also find many of those activities irritating and even oppressive. The government that confines itself to the fewest activities will coerce and irritate the fewest citizens. The government that undertakes to do everything, irritates and oppresses everyone. Oppression is defined in the minds of the oppressed. The sources of irritation and oppression vary. With time the irritation and oppression become the unifying forces in society. Though the citizens will not agree on which activities of government are evil, they will agree that the government is evil. When the citizens agree that government is evil, they turn against it. Government resorts to force and tyranny to preserve itself. We are already well down this road. Inevitably, an ever expanding government leads to despotism, which ultimately leads to total loss of support, and to collapse.

History shows that some coercion by government is as essential to maintaining civilization as are breaking and killing by the military. In our times many fail to recognize that coercion, even when necessary, is still evil. Just as with military force, remove the necessity and all that remains is evil. The only way we can save our government and our civilization is to back down the coercion. The mere fact that almost every function that we see as governmental has been performed privately, exposes the fallacy in the claim that government must do all these things.

Freedom and the American Dream will shrivel and die, unless we reduce the coercive activities of government to the bare minimum. Government must intervene only when the need is compelling – not just because intervention might, with luck, somehow do some good. The citizens must not feel oppressed by their government.

Government is often the tool of the lazy. Rather than making the effort to pursue their causes through the private sector, the lazy resort to the force of government. That force invariably corrupts even worthy causes. Accountability goes out the window. Then the causes have a way of losing their worthiness. Activities that enjoy overwhelming support are the easiest to pursue without government. Thus, the irony that those whose agendas are the least popular are the ones most in need of government coercion to carry out their schemes. The agendas of the lazy and the unpopular merge to become the agenda of government – the agenda of failure.

The more government coerces, the more it wants to coerce. Any government that does not control its appetite for coercion will ultimately succumb to its own toxin. Take no comfort that the coercive beast will destroy itself. The carcass of the fallen monster will crush us all. We must put government on a low coercion diet, rather than waiting for it to bloat and die.

There is only one way to keep government from being seen as an oppressor. We must limit government to activities supported by the overwhelming majority of the people. A simple majority vote to mandate intrusion into the lives of everyone, isn't enough. We must require super majorities of at least 60 percent to begin or continue any intrusion by government. This rule must apply to elected bodies and referendums. I have no patience with those who claim the requirement of super majorities will unduly hamper government. Six out of 10 still allows the overruling of many dissenters – perhaps too many. There is no magic in 60 percent. Experience may dictate the need for higher approval percentages. Different categories of actions may require different levels of support. Sixty percent is a good place to start. Simple majorities could still abolish programs.

Making super majority votes the rule, rather than the exception, will not be easy. The idea that half plus one controls has deep roots. We must ask the question, Why should 49 percent of us be forced to march to the drum of 51 percent? Not only that, Why should the 49 percent be forced to pay for a parade they don't even want? If only 51 percent can agree, Is it not far better that all be left free to pursue their goals through the private sector without coercion?

INDEX

Will your friends and relatives enjoy reading *Prostitutes, Margarine and Handguns* as much as you have? If copies are not available at stores in your area, you may order them from:

Pragmatic Publications, Dept. B
18440 29½ Mile Road
Springport, Michigan 49284

Please enclose check or money order for $9.00 (U.S.A.) for each copy, plus $2.00 per copy for packaging and shipping. Michigan residents must add 6% sales tax. There will be no charge for packaging and shipping for orders of 5 or more books shipped to the same address. Please provide the information shown on the sample order form below.

We will be pleased to quote quantity discounts and wholesale prices on request.

ORDER FORM

To: Pragmatic Publications, Dept. B
18440 29½ Mile Road
Springport, Michigan 49284

*Prostitutes, Margarine and Handguns*________ x $9.00 ________
(No. copies)

6% sales tax – $0.54 per copy (Mich. residents) ________

$2.00 per copy shipping (4 or less copies) . . . . . ________

Total enclosed . . ________

Please ship to:

____
(Name)

____
(Street & number)

____
(City, state and zip code)

ABOUT THE AUTHOR

Albert McCallum grew up in the farming country of Western Michigan where he graduated from Sparta High School in 1957. He earned a B.S. degree in civil engineering from Michigan State University, after which he served as an engineer for the Michigan State Highway Department, and held several other engineering positions. Al earned a Juris Doctor degree, *cum laude,* from the University of Michigan and has practiced law for over twenty five years. For twenty one years Al was with the Legal Department of Consumers Power Company, one of the nations largest utility companies, retiring as Senior Attorney and head of the contract section. Al also operated a farm, served as a school board president, and taught business law. He authors a weekly newspaper column, *Thoughts, Ramblings and Observations.* Al has three adult children and resides with his wife, Arlona, on their farm near Springport, Michigan where he practices law part time when he isn't writing.